BOTANICAL RESEARCH AND PRACTICES

TRANSGENIC PLANTS

RECENT DEVELOPMENTS

Botanical Research and Practices

Additional books in this series can be found on Nova's website under the Series tab.

Additional e-books in this series can be found on Nova's website under the e-book tab.

Genetics - Research and Issues

Additional books in this series can be found on Nova's website under the Series tab.

Additional e-books in this series can be found on Nova's website under the e-book tab.

BOTANICAL RESEARCH AND PRACTICES

TRANSGENIC PLANTS
RECENT DEVELOPMENTS

SHEN YAO ZHU
AND
JIANG LO HU
EDITORS

nova publishers
New York

For permission to use material from this book please contact us:
Telephone 631-231-7269; Fax 631-231-8175
Web Site: http://www.novapublishers.com

Library of Congress Cataloging-in-Publication Data

Transgenic plants : recent developments / editors: Shen Yao Zhu and Jiang Lo Hu.
p. cm.
Includes index.
ISBN 978-1-62257-245-8 (hardcover)
1. Transgenic plants. 2. Transgenic plants--Risk assessment. 3. Plant genetic engineering. I. Zhu, Shen Yao. II. Hu, Jiang Lo.
SB123.57.T735 2012
632'.8--dc23

2012021146

Published by Nova Science Publishers, Inc. † New York

CONTENTS

PREFACE

Plants whose DNA is modified using genetic engineering techniques are known as transgenic plants. In most cases the aim is to introduce a new trait to the plant which does not occur naturally in this species. Examples include resistance to certain pests, diseases or environmental conditions, or the production of a certain nutrient or pharmaceutical agent. This new book gathers and presents current research on transgenic plants including an examination of the release, persistence and transport of transgenic plant DNA in saturated and unsaturated mediums, such as soils and sediments; high-efficient plant cultivation with artificial light sources; the impact of three antibiotics on seed germination, hydroponic growth and nutritional qualities of vegetables; growth responses of Bt and non-Bt cottons to soil phosphorous, copper and cadmium levels; site-specific recombination systems for genome manipulation in transgenic plants; and amplification of small interfering RNAs in transgenic plants.

Chapter 1 - Purified bacterial cells and DNA, the signature of life, have been extensively studied in a wide range of environments and in different microbial ecosystems. On the contrary the release, persistence and transport of transgenic genes and the fate of transgenic DNA in different environmental compartments, have not been investigated in sufficient detail and depth, and are less well documented. Modelling DNA release from plant material may help to understand to the destiny of released plant DNA (transgenic or non transgenic) and its dispersion into the environment.

The question of the environmental impact and risk of transgene transfer from genetically modified plants (GMPs) to the natural environment remains a subject of intense scientific and also political debate. Thus, the questions concerning the fate of transgenes in the environment are not easy to answer, since the mechanisms and parameters controlling these processes are complex and not completely understood. The deliberate or accidental release of genetically engineered microorganisms or antibiotic resistance genes from transgenic plants into the environment may be a possible source of biological contamination of soil, surface and ground water. The main potential risk of the presence of recombinant DNA (rDNA) in the environment is the persistence of transgenes released from transgenic plants and their transport into different environmental compartments. Consequently, the ability of the transgenes to transform competent soil or aquatic microorganisms is a key question.

This chapter is part of a wider study examining the release, persistence and transport of transgenic plant DNA in saturated and unsaturated mediums, such as soils and sediments. The kinetics of plant leaf mass loss and DNA release as well as the long-term physical persistence and biological activity of transgenic genes such as *aadA* and *bla*TEM marker genes from transgenic plant DNA are developed. The data presented here suggest that plant leaf decomposition and DNA release may be described by a two-compartment first order model. This model indicates a potential for biologically active DNA to be transported over considerable distances in saturated and unsaturated medium. Additionally, the data demonstrate the persistence of DNA in the environment over long time periods and provide new insights into the fate of transgenic plant DNA in soil and aquatic environments. They also put an accent on the question regarding the possible ecological implications of the presence of extracellular transgenic DNA in groundwater.

Chapter 2 - Light is a vital environmental factor that affects plant growth and development by acting on plants not only as the energy source for photosynthesis, but also as an environmental signal for photomorphogenesis. Light requirements of plants are subjected to species, cultivar, growth and developmental stages, environmental conditions and manipulation target of yield and quality. Therefore, detailed studies on light quality requirements, i.e. light fomula (LF), based on biological and physiological requirement are urgently needed for getting high yield and good quality of protected plants. With the development of semiconductor solid light sources, light-emitting diodes (LEDs), detailed research on plant biology and physiology responses tomonochromatic light became possible. Therefore, the light quality requirements for all kinds of plants can be precisely and extensively studied to obtain the optimal light spectrum component for high productivity and good quality. The LF was defined as optimized monochromatic light component illuminated by LEDs for getting higher biomass and better nutritional quality of plants. Based on published literatures, red and blue light spectrum are macro-necessary LF components, and purple, green, yellow, cyan, orange and other visible light spectra are micro-beneficial LF components, while far-red and ultraviolet light spectrum is functional LF components. Generally, a kind of LF is composed of one or two kinds of necessary light spectra, several micro-beneficial and functional light spectra for special plant during a certain period and under certain environmental conditions. The LF is of crucial scientific significance that should be established for plants grown with artificial light. More importantly, the LF is an important part of light environment management strategy (LEMS). The LEMS refers to a comprehensive management method of light environment, including light intensity, LF and photoperiod for one special plant, which should be established for plants with artificial light sources. It is assumed that the LF and LEMS will be preferentially applied in plant factory with entire artificial light source.

Chapter 3 - The harvesting and storing of alfalfa is a routine practice in the agricultural industry worldwide. A study of the physiological and biochemical changes that occur after harvesting may help to understand how plants respond to this process.

To investigate gene expression in harvested alfalfa (*Medicago sativa*), cDNA from non-harvested and harvested plants in the field was subjected to subtractive hybridization to identify, in particular, those genes that are induced by the harvesting treatment. Three different genes, named*, hi7, hi11 and hi12,* were isolated and analysed. Northern blot analysis confirmed that *hi7, hi11* and *hi12* are strongly induced during a post harvest incubation period. The promoters of these genes were isolated and characterized. Reporter gene driven by the promoter has been transformed into transgenic plants. The hi12 promoter showed it is harvesting inducible.

Chapter 4 - Two experiments were conducted to investigate the impacts of three antibiotics on seed germination, hydroponic growth and nutritional qualities of vegetables. Seed germination test of three vegetables (cucumber, rape and Chinese cabbage) with six concentrations of tetracycline (TC), chlorotetracycline (CTC) and oxytetracycline (OTC) in form of hydrochlorides was carried out to study the impacts of these antibiotics on root elongation. Hydroponic experiment was designed to explore the growth inhibition and nutritional quality responses of lettuce grown in greenhouse with presence of various concentrations of CTC and OTC. The results showed that three antibiotics inhibited root elongation of germinated seeds, and the inhibitory rates depended on the drug species, antibiotic concentrations and vegetable species. Shoot and root biomass of lettuce were significantly reduced when antibiotics added in the nutrient solution. In addition, the soluble sugar content in leaves and petiole, and soluble protein in petiole were modified by antibiotic pollution. To conclude, antibiotic pollution of soil and water will adversely impact the seed germination and growth of protected vegetables grown in soil or hydroponics.

Chapter 5 - Two glasshouse experiments were conducted to investigate the growth differences of Bt- and non-Bt cotton on soils applied with various phosphorus (P), copper (Cu) and cadmium (Cd) levels. The results showed that high P level had no significant effect on shoot dry weight, root length and root dry weight of non-Bt cotton compared with low P treatment. However, high P level increased the shoot dry weight, root length, root dry weight and specific root length of Bt cotton. There was no difference between biological indices except specific root length between Bt cotton and non-Bt cotton under high P condition. Soil Cd and Cu addition inhibited the growth of both Bt and non-Bt cotton, with a significant decrease in shoot height, shoot dry weight and root dry weight. However, non-Bt cotton presented higher shoot height and shoot dry weight at the higher Cd and Cu levels.

Chapter 6 - By enabling the development of new crop varieties with increased nutritional content, disease resistance, and stress tolerance, plant biotechnology plays an important role in addressing the food needs of the world's population. According to ISAAA Brief 39-2008, the area of land used globally for cultivating transgenic crops grew from 114.3 million hectares in 2007 to 125 million hectares in 2008. In 2007, the direct global farm income benefit from biotech crops was $10.1 billion. So far, the majority of transgenic crops have been produced either by *Agrobacterium*-mediated transformation or by biolistic transformation. The gene controlling a trait of interest is usually delivered along with a selectable marker gene (such as an antibiotic-resistance gene) for later selection of the transgenic plants from the bulk of non-transformed plants. After serving this purpose, the selectable marker gene is no longer needed. However, this approach for producing transgenic crops has raised concerns from both the regulatory and public sectors over the presence of selectable marker genes in the food supply or their possible escape into the environment through the outcrossing of transgenic crops to related species. Therefore, research in plant transgenesis continues to focus on ways to generate transgenic crops free of selectable marker genes. Several approaches for producing selectable-marker-free transgenic crops have been reported. Among these approaches, one of the most powerful is the application of microbial site-specific recombination systems to the transformation technology. This article will discuss the promise of transgenic breeding for future agriculture, the public concerns over genetically modified plants, and the utilization of site-specific recombination systems (including the novel ParA system recently described by our laboratory) for the production of marker-free transgenic plants. The use of site-specific recombination systems to tailor controlled transgene expression in plants and to genetically engineer plants for breeding programs will be discussed, as well.

Chapter 7 - RNA interference (RNAi) is a rapidly popularized technology which contributes to progress in basic plant sciences and applied plant biotechnology. Most plant RNAi vectors transcribe an inverted repeat sequence consisting of target sequences. The resultant hairpin RNAs are processed into 21~25-nucleotide-long small interfering RNAs (siRNAs). These primary siRNAs are incorporated into the ARGONAUTE1-containing protein complexes, and then guide the sequence-specific cleavage of mRNAs. Primary siRNAs often trigger the synthesis of secondary siRNAs. In this amplification process, annealing of primary siRNAs to target mRNAs is followed by recruitment of RNA-DEPENDENT RNA POLYMERASE6 (RDR6), and then complementary RNA molecules against the target mRNAs are synthesized. The newly synthesized dsRNAs are processed into secondary siRNAs by redundant activities of DICER-LIKE4 and DICER-LIKE2. Since the target region of secondary siRNAs spreads into regions outside the primary target site, this phenomenon is called transitivity. Spreading of RNAi target via transitivity deteriorates specificity of RNAi. Therefore, elucidation of mechanistic issues of transitivity is important for researchers who use RNAi plants. Transitivity is frequently observed in transgenic plants expressing target transgene sequences under the presence of primary siRNAs, and there is only one report showing transitivity along the endogenous mRNA templates. It remains unclear why most endogenous transcripts seem to be an inefficient template for RDR6 during transitivity, but the limited transitivity along the endogenous transcripts apparently contributes to the maintenance of specificity of RNAi. This review is focused on transitivity observed in transgenic plants.

In: Transgenic Plants
Editors: Shen Y. Zhu and Jiang L.Hu
ISBN 978-1-62257-245-8

Chapter 1

PLANT LEAF DECOMPOSITION, DNA RELEASE, PERSISTENCE AND TRANSFER INTO THE ENVIRONMENT

John Poté* and Walter Wildi
University of Geneva, Institute F.A. Forel and Institute of Environmental Sciences, Versoix, Switzerland

ABSTRACT

Purified bacterial cells and DNA, the signature of life, have been extensively studied in a wide range of environments and in different microbial ecosystems. On the contrary the release, persistence and transport of transgenic genes and the fate of transgenic DNA in different environmental compartments, have not been investigated in sufficient detail and depth, and are less well documented. Modelling DNA release from plant material may help to understand to the destiny of released plant DNA (transgenic or non transgenic) and its dispersion into the environment.

The question of the environmental impact and risk of transgene transfer from genetically modified plants (GMPs) to the natural environment remains a subject of intense scientific and also political debate. Thus, the questions concerning the fate of transgenes in the environment are not easy to answer, since the mechanisms and

* Corresponding author. Tel.: +41-22-379-03-21; Fax: + 41 22 379 03 29. E-mail: john.pote@unige.ch.

parameters controlling these processes are complex and not completely understood. The deliberate or accidental release of genetically engineered microorganisms or antibiotic resistance genes from transgenic plants into the environment may be a possible source of biological contamination of soil, surface and ground water. The main potential risk of the presence of recombinant DNA (rDNA) in the environment is the persistence of transgenes released from transgenic plants and their transport into different environmental compartments. Consequently, the ability of the transgenes to transform competent soil or aquatic microorganisms is a key question.

This chapter is part of a wider study examining the release, persistence and transport of transgenic plant DNA in saturated and unsaturated mediums, such as soils and sediments. The kinetics of plant leaf mass loss and DNA release as well as the long-term physical persistence and biological activity of transgenic genes such as *aadA* and *bla*TEM marker genes from transgenic plant DNA are developed. The data presented here suggest that plant leaf decomposition and DNA release may be described by a two-compartment first order model. This model indicates a potential for biologically active DNA to be transported over considerable distances in saturated and unsaturated medium. Additionally, the data demonstrate the persistence of DNA in the environment over long time periods and provide new insights into the fate of transgenic plant DNA in soil and aquatic environments. They also put an accent on the question regarding the possible ecological implications of the presence of extracellular transgenic DNA in groundwater.

Keywords: Plant DNA; transgenic plants; DNA release; Antibiotic gene resistant; persistence; degradation; transport, exponential model

1. Introduction

The global agricultural surface cultivated by genetically modified plants (GMPs) continues to increase. In 2010, more than 120 million hectares of GMPs were sown world wide. Herbicide tolerance and microbial insect resistance properties are present in all of the major GMPs including maize, soybeans, rapeseed and cotton. These GMPs are grown mainly in the United States, Argentina, Canada and Brazil. The herbicide-tolerant soybeans account for about 78.6% of the world's biotech crops [1]. Technological innovations, such as GMPs, are considered bringing their own set of benefits and potential risks for the environment. Therefore, the new transgenic species incorporated in crops should be evaluated at least for environmental safety, food security and ethical aspects [2]. One of the main risks of transgenic plants is the release and dispersal of their transgenes, before they are degraded, through gene flow into the environment and into wild populations. Ecological effects may be caused by deliberate or accidental release of genetically engineered micro-organisms, first by survival, then by multiplication and dispersal of the inoculated cells as well as their transfer of recombinant DNA (rDNA) to indigenous microorganisms; these organisms have therefore to be monitored in the environment [3]. Additionally, the transgenic plants may spread their transgenes through gene flow into wild populations causing genetic pollution and giving rise to potential ecosystem disruption. The following issues associated with genes from transgenic plants must be dealt with: (i) emission, dispersal, and deposition of transgenic pollen, (ii) introgression of the transgene into wild species, (iii) stabilisation and spread of the transgene in wild species, (iv) ecological effects of the transgene in the new host population [4]. DNA from plants has been released for very long times into the environment naturally [5, 6]. According to different studies, the principal ways of transgene release from GMPs and their entry into the environment were summarised [7,8]: (i) by air transport (mainly through pollen dispersal), (ii) by decay in the residuesphere, *i.e.* enzymatic degradation of cell structures by plant pathogens and release of DNA, followed by infiltration into the soil and possibly into the groundwater, or runoff towards the surface water, (iii) by degradation of plant debris incorporated in the soil, followed by infiltration, or runoff towards the surface water, (iv) by degradation and infiltration from the rhizosphere (sloughing of root cap cells) into the soil, surface water and possibly to the groundwater.

Cell lyses during plant material decomposition can be considered as the main mechanism participating in the release of plant DNA to the environment [3, 9-12]. Studies regarding the assessment of the potential risk of the use of transgenic plants in agriculture are mostly focussed in the fate of released recombinant DNA into the environment. However, according to many scientific publications [8, 12, 13-20], rDNA released during plant material decomposition in the soil, sediments and surface water is generally poorly characterized. Also, little is known about the movement, temporal or spatial dispersal of rDNA in subsurface soil and fate of rDNA in the aquatic environment. It is thus important to focus further research efforts on these aspects.

Some studies performed on plant material decomposition demonstrated that (i) plant material decomposition in the soil is temporally limiting the transfer of DNA to the soil; before release, important quantities of DNA is degraded inside the plant tissues during decomposition [17], (ii) plant DNA release, long term persistence and the movement of DNA towards the vadose zone of the soil and to the groundwater [18, 19]. On the other hand, field data demonstrated that many of the genes such as *bla*TEM116 gene (encoding resistance to ampicillin, which belongs to the beta-lactam antibiotic family), and *aadA* genes (encoding resistance against to spectinomycin and streptomycin) are incorporated into GMPs. These genes are present naturally in the environment, as well as soil bacteria, and can be naturally resistant to a broad spectrum of antibiotics [12, 16, 20]. However, depending on environmental factors, further attention is needed with respect to these aspects concerning long-term cropping of GMPs. The assessment of the environmental impact of rDNA from GMPs, and in particular its kinetic aspects, requires the use of models capable of assessing the characteristics of the initial source, plant material decomposition processes, rDNA release processes, the means of transport and the interaction of rDNA with biotic and abiotic parameters of the environment [11, 17, 21].

2. Fate of DNA into the Environment

While DNA analysis has become central to work on natural gene exchange, forensic analyses, soil bioremediation, genetically modified organisms, exobiology, and palaeontology, fundamental questions about DNA resistance to degradation remain. The Earth is virtually entirely covered in nucleic acid as well as DNA from plants has been released into the environment naturally [6]. Thus, nucleic acids are ubiquitous in many environments, including fresh water, marine and surface water column and sediments, soil and terrestrial subsurface [9, 22, 23]. The fate of DNA in the environment can be summarized as follows: (i) DNA release from organisms (plant, microorganisms, animals), (ii) persistence of extracellular DNA according to environmental parameters; (iii) adsorption of extracellular DNA to the soil matrix; (iv) degradation of extracellular DNA by DNases; (v) extracellular DNA transforms competent soil microorganisms (binding of DNA onto soil components does not eliminate the ability of bound DNA to transform competent soil microorganisms [6, 12, 13]); (vi) probable dispersal and vertical movement of extracellular DNA in unsaturated soil medium; (vii) extracellular DNA used as nutrient source by soil and aquatic microorganisms. Although, data about the frequency or likelihood of each of these steps is available, the entire sequence is difficult to monitor in the field. During these last years, many studies have been performed in field and microcosms to describe these processes [8, 24-26].

After released from organisms in the environment compartments described above, eDNA can be rapidly adsorbed to soil or sediment particles. Numerous experiments have demonstrated the mechanisms and importance of DNA adsorption onto soil components such as sand particles, clays minerals, or humic compounds [9, 28-32]. These results suggest that free DNA is rapidly degraded in the environment, but the adsorption of DNA onto soil components retards DNA degradation and constitutes a major mechanism of DNA molecule persistence in soil. However, a noticeable aspect of the majority of these studies is that the quality and quantity of DNA in plant materials (during decomposition) before release into different environments should be investigated.

2.1. Fate of Extracellular DNA in Aquatic Environment

The type of aquatic environment affects the amount of DNA present and what can be recovered using various extraction methods [32]. The surface and marine water concentration of dissolved eDNA has been estimated to range from 2 to 90 $\mu g\ L^{-1}$ [27-31]. In the sediments from Lake Geneva, the eDNA concentration ranged from 15.3 to 22.4 $\mu g\ g^{-1}$ [21]. Extracellular DNA can be used as C, N and P sources by heterotrophic microorganisms and plays a significant role in bacterial biofilm formation in the sediments. It is therefore considered as a source of nutrient and genetic information in aquatic environment [26]. It has been demonstrated that eDNA can be adsorbed onto sediment components, escapes from enzymatic degradation or chemical destruction and play a part important in horizontal gene transfer (HGT) [30, 32, 33-35]. The fate of eDNA in groundwater aquifer is poorly investigated. The concentration of eDNA in Geneva groundwater range from 5.3 to 7.8 $\mu g\ L^{-1}$ [19]. Numerous experiments have demonstrated the mechanisms and importance of DNA adsorption onto sediment components such as sand particles, clays minerals, or humic compounds. These results suggest that free DNA is rapidly degraded in the environment, but the adsorption of DNA onto sediments components retards DNA degradation and constitutes a major mechanism of DNA molecule persistence in the sediments. However, noticeable aspects of the majority of these studies are that the quality and quantity of rDNA in plant materials (during decomposition) before release, the potentiality of rDNA (after released in both sediments and water column) to transform competent bacteria in the aquatic compartments are lacking.

2.2. Persistence of Transgenic DNA in Aquatic Environment

The processes resulting from transgenic plants in agriculture fields such; pollen dispersal, decomposition of plant material in soil and DNA release, water flow and rDNA transport in vadose zone, presence of transgenic plant crops near surface waters, the input of crop byproducts in the streams, had promoted studies on the fate of transgenic DNA in the aquatic environment. These studies are summarised in the many literatures [13, 14, 19, 21, 26, 33]. It has been demonstrated that DNA from transgenic plant can persist in aquatic environment and increases the probability of horizontal gene transfer in microorganisms by transduction, transformation and conjugation. But according to the water characteristics, important quantities of DNA are degraded rapidly in water column. Bacteria transformation can only occur if the bacteria cell is in a particular physiological state named competence. The development of competence is usually monitored by detecting bacterial transformation, which involves binding of exogenous DNA on the external surface of the bacterial cell, DNA uptake usually as a single strand, insertion of the single strand in the chromosome or double strand plasmid reconstruction and gene expression [26]. Thus, the studies performed using extracellular purified rDNA (plasmid and chromosomal transgenic DNA) in microcosms to monitor its persistence and transformability in water concluded that no antibiotic or bacteria transformation were occurred under their experimentation conditions [32-35]. Some studies have been performed in field to monitor the persistence of rDNA from transgenic Bt corn (*cry1 Ab* gene) into aquatic environment [13, 14]. Their data suggest the persistence of transgenes in water and sediments as well as its potential impacts to aquatic leaving organisms. These researches confirmed that transgene expression and bacteria antibiotic resistance and transformability in water column and sediments are not sufficiently investigated to confirm many of the hypotheses. Therefore, they propose that further research and monitoring efforts on the identity of rDNA in aquatic environment are required.

3. Plant Material Decomposition and DNA Release

Plant materials are one of the principal source inputs of organic nutrients such as carbon, nitrogen, and phosphorus in aquatic environment. Prediction of plant material decomposition in the environment is correlated to the release of macromolecule content including cellulose, holocellulose, lignin, tannins and DNA [17, 33, 34]. The decomposition of plant material in an ecological system can be determined using a variety of methods including mass loss, variation in chemical composition, changes in C/N ratio, CO_2 emission, and release of molecules and elements [38-40].

The model commonly used to describe the decomposition of plant litter in soil is composed of the single (Eq. 1) and double exponential (Eq. 2) decay functions [41-43]:

$$M_{\mathrm{DRY}}(t) = M_{\mathrm{DRY0}}\mathrm{e}^{-kDRY \cdot t} \tag{1}$$

$$M_{\mathrm{DRY}}(t) = A\mathrm{e}^{-ka \cdot t} + B\mathrm{e}^{-kb \cdot t} \tag{2}$$

where $M_{DRY}(t)$ is the dry matter weight at time t, M_{DRY0} is the initial dry matter weight, and k_{DRY} is the loss rate constant (decomposition rate), A and B are the initial dry matter proportions of the fast and the slow decomposing compartments, respectively ($M_{\mathrm{DRY0}} = A + B$), and ka and kb are respectively the decay rate constants for A and B fractions. Each fraction decreases exponentially.

The half-decomposition time of the dry matter is given by:

$$T_{DRY\,1/2} = \frac{1}{k_{DRY}} \ln 2 \tag{3}$$

Several factors such as structural composition of leaves, microbial activity, invertebrates, as well as temperature and humidity play a key role in the decomposition of plant materials in the environment.

It has been demonstrated that the DNA mass loss concentration is not constant in the dry matter of leaves, showing decrease through time [15, 17]. Therefore, DNA weight can be determined from the measurement of the DNA mass concentration in leaves, $C_{DNA}(t)$, and the previous dry matter weight of samples $M_{DRY}(t)$, using the direct relation:

$$M_{DNA}(t) = C_{DNA}(t) \cdot M_{DRY}(t) \quad (4)$$

This equation (Eq. 4), an exponential model fitting, allows the determination of both k_{DNA} and the associated DNA half-decrease time given by:

$$T_{DNA}1/2 = \frac{1}{k_{DNA}} \cdot \ln 2 \quad (5)$$

Interests in the fate of recombinant DNA in the environment, many studies have been performed in field and microcosms on the decomposition of plant material and DNA release in the soil [10, 11, 17, 44-46]. The results indicate that the decrease of dry matter in leaves in both varieties was better described by a two-compartment (double exponential) first order model. Flores et al. [47] compared the decomposition of various transgenic *Bt* and non-*Bt* plants. The authors concluded that the biomass of transgenic *Bt* plant was decomposed less in soil than the biomass of their near-isogenic non- *Bt* plant counterparts. The authors recommended others studies to clarify the environmental impacts of lower degradation of the biomass of *Bt* plant; Rosi-Marshall et al. [13] measured breakdown rates of *Bt* and non-*Bt* corn litter to determine whether the *Bt* δ-endo-toxin influences rates of organic matter processing in their study stream. They found that no difference in decomposition rates between *Bt* and non-*Bt* corn suggesting that transgenic and nontransgenic corn byproducts are processed in the same manner in stream ecosystems; Levy-Booth et al. [15] monitored the decomposition of roundup ready leaf biomass in the soil microcosms. The soybean biomass decomposition was described using a single-phase exponential equation (eq. 1). The authors founded that the biomass of roundup ready soybean leaves was 8.6% less than non-transgenic soybean leaves after 30 days in the soil microcosms, and concluded that their study was not an investigation of DNA entry pathways under field conditions, but a model system to investigate the entry of DNA in a controlled environment. The data concerning the decrease of transgenic DNA in decomposing plant, monitoring and quantifying the transgenes in aquatic environment remain lacking. Some published data [21] performed in field and microcosms to investigate these aspects in aquatic environment have been performed using non transgenic tomato leaves. For the perspectives, the obtained results recommended the use of transgenic plants.

Further extension and application of these results should be performed with specific transgenes. The possibility of bacteria antibiotic resistance and the horizontal gene transfer to sediment microorganisms is also worthy of investigation. The relative contribution of various plant communities (e.g. sediment bacteria or endo/exophytes) in facilitating extra versus intracellular plant DNA degradation also needs to be examined. Fulfilling of data could provide on the examination of both the quantitative and qualitative aspects of rDNA in plant material decomposition in the water column and sediments. Quantitative measurement includes rDNA concentration in planta and in water and sediments after released. Qualitative measurements include both the degradation of rDNA and its potential to transform bacteria leading to the production of antibiotic resistant bacteria.

4. Water Flow and DNA Transport in Subsurface Environment

The unsaturated zone is inextricably involved in many aspects of hydrology: infiltration, vertical advection by water capillarity, evaporation, groundwater recharge, soil moisture storage, and soil erosion [48]. It also contributes to the spatial and temporal distributions of plant communities under naturally occurring rainfed conditions and serves as a modifying influence on the production of cultivated crop species.

Few studies have investigated the simultaneous processes of plant material degradation, transgenic DNA release and transport by water in unsaturated and saturated zones, and reach aquatic environment such as groundwater, water column and sediments. Poté et al. [22] used extracted and purified DNA (plasmid pLEPO1 which confers resistence to spectinomycin and streptomycin) to study the transport of antibiotic genes in saturated and unsaturated soil medium using the laboratory columns. Contact between bacteria and the exogenous DNA was improved by mixing the soil and DNA in a microcosm with the result that bacteria were able to incorporate DNA in soil. These results suggest a potential for biologically active DNA to be transported over considerable distances in saturated soil and ground water. Ceccherini et al [24] used unsaturated soil column to study the vertical advection of 35S-nptII by water capillarity. The results show the vertical movement of recombinant DNA due the capillary rise and suggest the possibility of DNA degradation within the soil column. The results of these laboratory studies suggest that when extracellular DNA enters the soil environment, it might be dispersed by rain, irrigation-induced percolation, runoff or groundwater before being completely degraded. Gulden et al. [25] were able to quantify the amount of corn and soybean DNA released by plant roots into leachate water during growth and early plant decomposition in water. They showed that detectable quantities of plant target DNA were released into the soil environment and were moved by water during growth and early decomposition of roots. Poté et al. [18] used non transgenic tomatoes to simulate the natural processes of DNA release and vertical movement of released DNA in vadose zone using unsaturated soil column. The results suggest that DNA could be transported in subsurface soil and can reach the groundwater.

Water flow and rDNA dispersal in the environment concerns both the processes of transgenes release from GMPs and the evolution of active bacteria capable of incorporating this DNA into their genomes. Many literatures report on the widespread occurrence of plant DNA (non transgenic plant) in groundwater and traditional drinking water fountains in the Geneva Champagne Basin [19]. The observation supports the notion of plant DNA release, long term persistence and movement in the unsaturated medium as well as in groundwater aquifers.

5. Antibiotic Resistance Bacteria and Horizontal Gene Transfer

The antibiotic resistance bacteria (ARB), the transfer frequency of transgenes from plants to environmental bacteria and expression of these genes in these new bacterial hosts are considered as crucial issues when evaluating the impact of transgenic plants. Several studies indicate that such transfers are possible, and that these transfers are directly related to the transgenic status of these plants. Plants subjected to infection by pathogens are efficiently colonized by other soil micro-organisms, including some bacteria that develop genetic competence and are transformed by exogenous DNA including this released by the transgenic plant [49, 50]. The same group showed that a decaying plant suspension (residuesphere) was another hot spot for gene transfer between transgenic plants and colonizing soil bacteria [51]. When decaying, plant releases DNA in contact to metabolically active bacteria due to the presence of nutriments that permits bacterial growth and development of a genetic competence stage. However, plant-bacteria DNA transfer was detected only when the recipient bacteria contained the plant DNA regions flanking the marker gene on which homologous recombination could be initiated, indicating that the main barrier to gene transfer is not physical or physiological but genetic [49, 51-52].

For years, gene transfer between transgenic plants and bacteria was considered as theoretically possible but the lack of experimental confirmation led some scientists to conclude that such events are very unlikely. Consequently, several questions remain regarding gene transfer from plants to bacteria: (i) Can any prokaryote sequence cloned in the plant genome able to be transferred to any soil or sediment bacteria? Can these plant sequences be transferred at the same frequency or does the transfer frequency depend on DNA sequences prevalence on which homologous recombination can occur among soil and sediment bacteria? (ii) Can these prokaryote sequences be involved in the co-transfer of flanking DNA regions in the plant genome and what is the frequency of such specific events? (iii) Can natural genetic transformation the only mechanism that soil bacteria use to acquire transgene genes? (iv) Are transgenic plants-bacteria DNA transfer events limited to some environmental "hot spots" or do they also happen in the bulk soil? (v) What is the occurrence frequency of these plant-to-bacteria transfer events in these various environments?

These questions require the development of specific tools to discriminate between independent transfer events and clonal multiplication of initial transformants by the researches under *in vitro* conditions or/and in the field. Specific technical improvements are also necessary to determine if transfer events can occur under natural conditions requiring detecting specific transgene molecular signatures in bacteria genomes recovered from soil samples collected in a field planted with transgenic plants.

Another important question is related to the nucleotide similarity level that would potentially exist between plant transgene (donor DNA) and soil bacteria genomes (recipient organisms), a criterion susceptible to strongly regulate recombination mediated transgene gene integration in bacteria. For any plant transgene whose nucleotide sequence is available in public database experimental and *in silico* studies would be necessary to determine the actual potential of these transgenes to be fully or partly transferred to bacteria. These include soil or sediment metagenome sequence analysis to detect bacteria sequences similar to those of the transgenes and transformation tests in model bacteria to assess ability of these bacterial sequences to promote recombination with plant transgene genes. The studies should not be restricted to typical "agronomic" transgenes (cry, pat) but should also include those whose genes code for bio-active pharmaceutical molecules considering that their potential transfer to soil or sediment bacteria could have a stronger environmental or sanitary impact than agronomic transgenes.

6. A Brief Protocol to Study the Fate of rDNA in to the Aquatic Environment

The experiments for studying the fate of transgenic DNA into the aquatic environment can be performed in the microcosms (Figure 1) and different analyses can be performed as summarized in Figure 2. The study consists in 3 phases; (i) Decrease of transgenic DNA in planta during plant material decomposing in sediments, (ii) persistence of transgenic DNA in water column and sediments, and (iii) antibiotic-resistant bacteria (ARB) and horizontal gene transfer (HGT).

These microcosms present several advantages:

- No input relative pollution into the lake when using different substances
- the relatively large size of the equipments and mass of associated environmental media minimises the effect of small scale variation in the system
- by maintaining a natural ecosystem in the microcosms both biological and chemical interactions between water and sediment can be studied
- maintaining and stabilising the physicochemical parameters, pH, temperature, conductivity,.... of water at the fixed values throughout the experimentation mimics the environment
- recuperation and treatment of contaminated leachates discharged

6.1. Decrease of Transgenic DNA in Planta during Plant Material Decomposing in Sediments

Plant Leaf Mass Loss

According to our preliminary tests using non transgenic tobacco, the experimentation for plant material mass loss can be conducted for a total of 40 days (complete degradation of leaves).

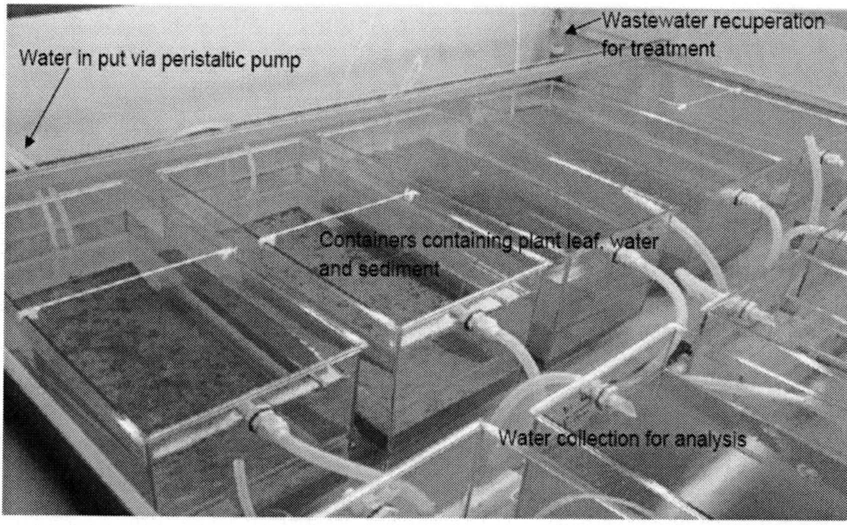

Figure 1. Simplified picture of platform microcosm developed at Forel Institute, University of Geneva to simulate lake conditions for studying the fate of pollutants in aquatic environment.

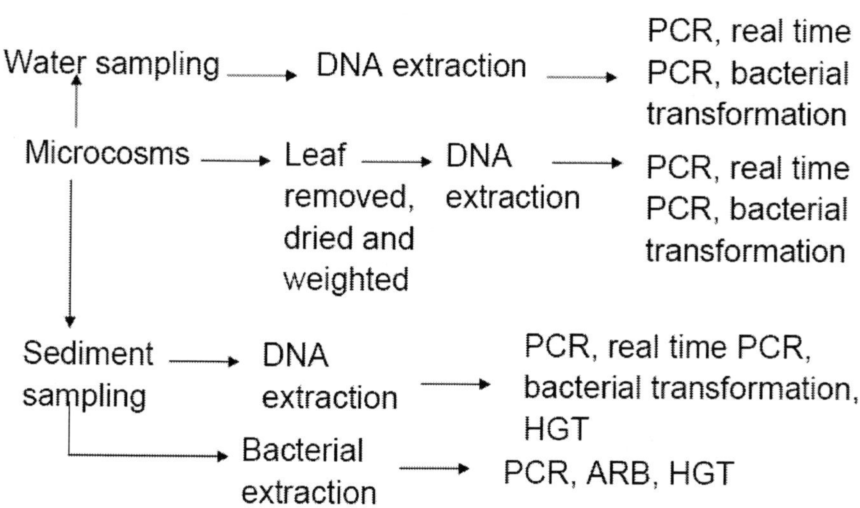

Figure 2. Simplified protocol for different analysis steps. The experiment can be conducted in triplicate in each set of conditions (three replicates by sampling point). The sediment can be sampled after complete degradation of leaves. Water has to be sampling in continuous-flow.

Leaves can be taken from the microcosms (Figure 1) after e.g. 1, 2, 5, 15, 25, 30 and 40 days (each container at sampling point with three replicates). Tobacco leaves can be removed from the microcosms and the nets dried at 30 °C for 5 days, and then weighed (Figure 2). Care has to be taken to remove all pieces of leaves attached to the net [21]. The results will be expressed as the mean of the triplicates with a standard deviation, and fitting models can be performed using SigmaPlot-9. The decomposition rate (modelling plant biomass) can be determined by fitting an exponential model to dry weight measurements of biomass loss and fitting this data to the decay time equations (Eq. 1-3).

Identity of Transgene in Planta

To determine the identity of transgene in planta during transgenic plant material decomposing in sediments, the quantitative and qualitative aspects of transgene have be determined in the same different sampling points as described above. Plant DNA will be extracted from the same samples of plant material sampled from sediment microcosms (Figure 1). The quantitative aspects of transgene include the concentration of total DNA and the number of transgene copies can be performed using real time PCR. Qualitative aspects include the relative degradation of transgene in agarose gel and the transformation frequencies indigenous bacteria such as for *E. Coli* DH 10B and *Acinetobacter sp.* BD 413 [12, 22, 50].

The identity of transgenes can be monitored using linear regression of the mean gene copies/g leaf biomass over time: $G = e^{-kt+ [gene]t=0}$, Where G: gene copies/ g dry plant material biomass, k: degradation coefficient, t: time in days and $[gene]_{t=0}$ gene copies/plant biomass at the time of incubation. The decrease of transgene in plant material decomposing in sediments can be modelling using Eq. 4 and 5.

6.2. Persistence of Transgenic DNA in Water Column and Sediments

For this task, two series of microcosm experimentations can be performed. The first consists to introduce 10 to 15 g of dry transgenic leaves in water column. These microcosms will serve to monitor the identity of released rDNA in water column. The leachates will be collected in continuous-flow in sterile polyethylene bottles. The analysis can be performed after e.g. 10, 20, 40, 60, 80 and 100 days, which correspond to 5-10 L of leachate water per sampling time. Before analyses (e.g. DNA extraction), water samples have to be concentrated more than 200-1000 times using per example Kühner Lyophilisateur at – 45°C and by Speed Vac plus SC 110A. These operations can not affect the quantity and the quality of DNA [19, 28]. DNA from concentrated water will be extracted and purified following the procedure described previously [18, 19, 27, 28]. For the second experimentation, the microcosms will be performed as described above §5.1. After 40 days (complete degradation of leaves), sediments can be sampled in several points of microcosms (by removing delicately the top layer of sediment 0-1 cm) as described in our previous studies [21].

All extracted DNA will therefore serve to monitor the quantitative and qualitative aspects of transgene in water column and sediments [12, 16, 22].

6.3. Antibiotic-resistant Bacteria and Horizontal Gene Transfer

The use of antibiotic resistance genes in the development of GMPs has given rise to environmental concern. Knowledge about the prevalence and diversity of the antibiotic resistance gene sediment bacteria communities is required to evaluate the possibility and ecological consequences of the transfer of these genes from GMPs to sediment bacteria. Our recent studies demonstrated that plant transgene genes that would have been naturally transferred into soil representatives under field conditions cannot be easily detected in soil bacteria because of the low frequency of transfer occurrence in nature, of the impossibility of extracting bacteria and DNA from large samples of soil and of cultivating *in vitro* more than a limited proportion of these bacteria. For example, the aadA gene conferring resistance to spectinomycin and streptomycin is the transgenes commonly used in transplastomic genetically modified plant constructions and the most common beta-lactamases are the TEM beta-lactamases that are encoded by the blaTEM1 gene and its descendants. The investigation of bacteria resistant against these antibiotics and the possibility of transgenes from GMPs transfer to bacteria has been largely investigated in our previous studies [20, 24]. According to the studies performed in the field soil conditions and microcosms, the development of methodologies for the detection of such events is complex and remains very difficult. However, these approaches are fundamentals and essentials to the evaluation of potential gene transfer and ecological consequences of transgene from GMPs into the environment [10, 11, 20, 30].

Conclusion

There is a paucity of information regarding the release and identity of transgenic DNA and its biological impacts in aquatic environment. The interdisciplinary approaches, e.g. rDNA release, persistence and bacteria antibiotic resistance and HGT, and their correlation with sediment and water characteristics are very important and can be of high interest to the international scientific community. Numerous questions remain about the role that natural environments play in the maintenance and dispersion of antibiotic resistance microorganisms and about the frequency with which the genes are exchanged among indigenous bacteria in aquatic environments.

References

[1] James, C., 2007. Global status of commercialized biotech/GM crops: 2007. In: ISAAA Brief No. 37. International Service for the Acquisition of Agri-Biotech Applications, Ithaca, NY, USA.
[2] Singh, O.V., Ghai, S., Paul, D., Jain, R.K., 2006. Genetically modified crops: success, safety assessment, and public concern. *Appl. Microbiol. Biotechn.* 71, 598–607.
[3] Paget, E., Simonet, P., 1994. On the track of natural transformation in soil. *FEMS Microbiol. Ecol.* 15, 109-118.
[4] Evenhuis, A., Zadok, J.C., 1991. Possible hazards to wild plants of growing transgenic plants. A contribution to risk analysis. *Euphytica* 55, 81–84.
[5] Soltis, P.S., Soltis, D.E., Smiley, C.J., 1992. An rbcl sequence from a Miocene Taxodium (bald cypresse). *Proc. Natl. Acad. Sci. USA* 89, 449–451.
[6] Trevors, J.T., 1996. Nucleic acids in the environment. *Current Opin. Biotech.* 7, 331–336.
[7] Poté, J. Devenir de l'ADN d'origine végétale dans les compartiments environnementaux: analyse systémique, rémanence et transport de l'ADN transgénique dans le sol. Ph.D thesis. *Terre et Environnement* 46, 200pp.
[8] Levy-Booth, D.J., Campbell, R.G., Gulden, R.H., et al., 2007. Cycling of extracellular DNA in the soil environment. *Soil Biology and Biochemistry* 39, 2977-2991.

[9] Blum, S.A.E., Lorenz, M.G., Wackernagel, W., 1997. Mechanism of retarded DNA degradation and prokaryotic origin of DNases in nonsterile soils. *Systematic and Applied Microbiology* 20, 513-521.

[10] Paget, E., Lebrun, M., Freyssinet, G., Simonet, P., 1998. The fate of recombinant plant DNA in soil. *European Journal of Soil Biology* 34, 81–84.

[11] Dale, P.J., Clarke, B., Fontes, E.M.G., 2002. Potential for the environmental impact of transgenic crops. *Nature Biotechnology* 20, 567-574.

[12] Ceccherini, M.T., Poté, J., Kay, E., Van, V.T., Maréchal, J., Pietramellara, G., Nannipieri, P., Vogel, T.M., Simonet, P., 2003. Degradation and Transformability of DNA from Transgenic Leaves. *Applied and Environmental Microbiology* 69, 673-678.

[13] Rosi-Marshall, E.J., Tank, J.L., Royer, T.V., Whiles, M.R., Evans-White, M., Chambers, C., Griffiths, N.A., Pokelsek, J., Stephen, M.L., 2007. Toxins in transgenic crop byproducts may affect headwater stream ecosystems. *Proceedings of the National Academy of Sciences* 104, 16204-16208.

[14] Douville, M., Gagné, F., Blaise, C., André, C., 2007. Occurrence and persistence of *Bacillus thuringiensis* (Bt) and transgenic Bt corn *cry1Ab* gene from an aquatic environment. *Ecotoxicol. Environ. Saf.* 66, 195–203.

[15] Levy-Booth, D.J., Campbell, R.G., Gulden, R.H., et al., 2008. Real-Time Polymerase Chain Reaction Monitoring of Recombinant DNA Entry into Soil from Decomposing Roundup Ready Leaf Biomass. *Journal of Agricultural and Food Chemistry* 56, 6339-6347.

[16] Pontiroli, A., Ceccherini, M.-T., Poté, J., Wildi, W., Kay, E., Nannipieri, P., Vogel, T., Simonet, P., Monier, J.-M., 2010. Long-term persistence and bacterial transformation potential of transplastomic plant DNA in soil, *Research in Microbiology*, Vol. 161, 326-334.

[17] Poté, J., Rossé, P., Rosselli, W., Tran Van., V., Wildi., W., 2005. Kinetics of mass and DNA decomposition in tomato leaves. *Chemosphere*, 61, 677-684.

[18] Poté, J., Rosselli, W., Wigger, A., Wildi, W., 2007. Release and leaching of plant DNA in unsaturated soil column. *Ecotox. Environ. Safety* 293-298.

[19] Poté, J., Mavingui, P., Navarro, I., Rosselli, W., Wildi, W., Simonet, P., Vogel, T.M., Extracellular Plant DNA in Geneva Groundwater and Traditional Artesian Drinking Water Fountains. *Chemosphere*, 75, 498-504.

[20] Demanèche, S., Sanguin, H., Poté, J., Navarro, E., Bernillon, D., Mavingui, P., Wildi, W., Vogel, T-M., Simonet, P. 2008. Antibiotic-resistant soil bacteria in transgenic plant fields. *PNAS*. 105, 3957-3962.

[21] Poté, J., Ackermann, A., Wildi, W., Plant leaf mass loss and DNA release in fresh water sediments. *Ecotox. Environ. Safety*, 1378-1383.

[22] Poté, J., Ceccherini, M.T., Tran Van, V., Rosselli, W., Wildi, W., Simonet, P., Vogel, T.M., 2003. Fate and Transport of Antibiotic Resistance Genes in Saturated Soil Columns. *Eur. J. Soil Biol.* 39, 65-71.

[23] Nielsen, D.R., Genucheten, M.TH., Biggar, J.W., 1986. Water flow and solute transport processes in unsaturated zone. *Wat. Res. Res.* 22, 89S-108S.

[24] Ceccherini, M.T., Ascher, J., Pietramellara, G., Vogel, T.M., Nannipieri, P., 2007. Vertical advection of extracellular DNA by water capillarity in soil columns. *Soil Biol. Biochem.* 39, 158–163.

[25] Gulden, R.H., Lerat, S., Hart, M.M., Powell, J.R., Trevors, J.T., Pauls, K.P., Klironomos, J.N., Swanton, C.J., 2005. Quantitation of Transgenic Plant DNA in Leachate Water: Real-Time Polymerase Chain Reaction Analysis. *J. Agric. Food Chem.* 53, 5858–5865.

[26] Pietramellara, G., Asher, J., Borgogni, F., Ceccherini, M.T., Guerri, G., Nannipieri, P. 2009. Extracellular DNA in soil and sediment: fate and ecological relevance. *Biol. Fertil. Soils*, 45, 219-235.

[27] Paul, J.H., Jeffrey, W.H., DeFlaun, M.F., 1987. Dynamics of extracellular DNA in marine environment. *Appl. Environ. Microbiol.* 53, 170-179.

[28] DeFlaun, M.F., Paul, J.H., Jeffrey, W.H., 1987. Distribution and molecular weight of dissolved DNA in subtropical estuarine and oceanic environments. *Mar. Ecol. Progr. ser.* 38, 65-73.

[29] Frostegärd, A., Courtois, S., Ramisse, V., Clerc S., Bernillon, D., Le Gall F., Jeannin, P., Nesme, X., Simonet, P., 1999. Quantification of Bias Related to the Extraction of DNA Directly from Soils. *Appl. Environ. Microbiol.* 65, 5409-5420.

[30] Lorenz, M.G., Wackernagel, W., 1994. Bacterial gene transfer by natural genetic transformation in the environment. *Microbiol. Rev.* 58, 563-602.

[31] Ogram, A., Sayler, G.S, Wackernagel, W., 1987. The extraction and purification of microbial DNA from sediments. *J. Microbiol. Meth.* 7, 57-66.

[32] England, L.S., Pollok, J., Vincent, M., Kreutzweiser, D., Fick, W., Trevors, J.T., Holmes, S.B., 2005. Persistence of extracellular baculoviral DNA in aquatic microcosms: extraction, purification and amplification by the polymerase chain reaction (PCR). *Mol. Cell. Probes*, 19, 75-80.

[33] Zhu, B., 2006. Degradation of plasmid and plant DNA in water microcosms monitored by natural transformation and real-time polymerase chain reaction (PCR). *Water Res.* 40,

[34] Aardema, B. W., Lorenz, M.G., Krumbein, W. E., 1983. Protection of Sediment-Adsorbed Transforming DNA Against Enzymatic Inactivation. *Appl. Environ. Microbiol.* 46, 417-420.

[35] Dale, P.J., Clarke, B., Fontes, E.M.G., 2002. Potential for the environmental impact of transgenic crops. *Nat. Biotechnol.* 20, 567-574.

[36] Kögel-Knabner, I., 2002. The macromolecular organic composition of plant and microbial residues as inputs to soil organic matter. *Soil Biology and Biochemistry* 34, 139–162.

[37] Coûteaux, M.M., Bottner, P., Berg, B., 1995. Litter decomposition climate and litter quality. *Tree* 10, 63-66.

[38] Monties, B., 1991. Plant cell walls as fibrous lignocellulosic composites: relations with lignin structure and function. *Animal feed Science and Technology* 32, 159–175.

[39] Coûteaux, M.M., Bottner, P., Berg, B., 1995. Litter decomposition climate and litter quality. *Tree* 10, 63-66.

[40] Eklind, Y., Kirchmann, H., 2000. Composting and storage of organic household waste with different litter amendments. II: nitrogen turnover and losses. *Bioresource Technolology* 74, 125-133.

[41] Jenny, H., Gell, S.P., Bingham, F.T., 1949. Comparative study of decomposition rates of organic matter in temperate and tropical regions. *Soil Science* 68, 419–432.

[42] Olson, J.S., 1963. Energy storage and the balance of producers and decomposers in ecological systems. *Ecology* 44, 322–331.

[43] Wieder, R.K., Lang, G.E., 1982. A critique of the analytical methods used in examining decomposition data obtained from litter bags. *Ecology* 63, 1636–1642.

[44] Widmer,F., Seidler, R.J., Watrud, L.S., 1996. Sensitive of transgenic plant marker gene persistence in soil microcosms. *Mol. Ecol.* 5, 603–613.

[45] Widmer, F., Seidler, R.J., Donegan, K.K., Reed, G.L., 1997. Quantification of transgenic plant marker gene persistence in the field. *Mol. Ecol.* 6, 1-7.

[46] Gebhard, F., Smalla, K., 1999. Monitoring field releases of transgenic modified sugar beets for persistence of transgenic plant DNA and horizontal gene transfer. *FEMS Microbiol. Ecol.* 28, 261–272.

[47] Flores, S., Saxena, D., Stotzky, G., 2005. Transgenic *Bt* plants decompose less in soil than non- *Bt* plants. *Soil Biol. Biochem.* 37, 1073-1082.

[48] Nielsen, D.R., Genucheten, M.TH., Biggar, J.W., 1986. Water flow and solute transport processes in unsaturated zone. *Wat. Res. Res.* 22, 89S-108S.

[49] Kay, E., Bertolla, F., Vogel, T.M., Simonet, P. 2002. Opportunistic Colonization of *Ralstonia solanacearum*-Infected Plants by *Acinetobacter* sp. and Its Natural Competence Development. *Microb. Ecol.* 43, 291-297.

[50] Kay, E., Vogel, T. M., Bertolla, F., Nalin, R., Simonet, P. 2002. In situ transfer of antibiotic resistance genes from transgenic (transplastomic) tobacco plants to bacteria. *Appl. Environ. Microbiol.* 68, 3345-3351.

[51] Rizzi, A., Pontiroli, A., Brusetti, L., Borin, S., Sorlini, C., Abruzzese, A., Sacchi, G. A., Vogel, T. M., Simonet, P., Bazzicalupo, M., Nielsen, K. M., Monier, J. M., Daffonchio, D. 2008. Strategy for in situ detection of natural transformation-based horizontal gene transfer events. *Appl. Environ. Microbiol.* 74, 1250-1254.

[52] Bertolla, F., Pepin, R., Passelegue-Robe, E., Paget, E., Simkin, A., Nesme, X., Simonet, P. 2000. Plant Genome Complexity May Be a Factor Limiting In Situ the Transfer of Transgenic Plant Genes to the Phytopathogen *Ralstonia solanacearum*. *Appl. Environ. Microbiol.* 66, 4161-4167.

In: Transgenic Plants
Editors: Shen Y. Zhu and Jiang L.Hu
ISBN 978-1-62257-245-8

Chapter 2

ESTABLISHMENT OF LIGHT FORMULA AND LIGHT ENVIRONMENTAL MANAGEMENT STRATEGY FOR HIGH-EFFICIENT PLANT CULTIVATION WITH ARTIFICIAL LIGHT SOURCES

Wenke Liu[1,2], Qichang Yang*[1,2], Lingling Wei[2] and Ruifeng Cheng[1,2]

[1]Institute of Environment and Sustainable Development in Agriculture, Chinese Academy of Agricultural Sciences, Beijing, China

[2]Key Lab. of Energy Conservation and Waste Management of Agricultural Structures, Ministry of Agriculture, Beijing, China

ABSTRACT

Light is a vital environmental factor that affects plant growth and development by acting on plants not only as the energy source for photosynthesis, but also as an environmental signal for photomorphogenesis. Light requirements of plants are subjected to species, cultivar, growth and developmental stages, environmental conditions and manipulation target of yield and quality. Therefore, detailed studies on light quality requirements, i.e. light fomula (LF),

* Corresponding author: Qichang Yang, Email: yangqichang1@vip.sina.com.

based on biological and physiological requirement are urgently needed for getting high yield and good quality of protected plants. With the development of semiconductor solid light sources, light-emitting diodes (LEDs), detailed research on plant biology and physiology responses tomonochromatic light became possible. Therefore, the light quality requirements for all kinds of plants can be precisely and extensively studied to obtain the optimal light spectrum component for high productivity and good quality. The LF was defined as optimized monochromatic light component illuminated by LEDs for getting higher biomass and better nutritional quality of plants. Based on published literatures, red and blue light spectrum are macro-necessary LF components, and purple, green, yellow, cyan, orange and other visible light spectra are micro-beneficial LF components, while far-red and ultraviolet light spectrum is functional LF components. Generally, a kind of LF is composed of one or two kinds of necessary light spectra, several micro-beneficial and functional light spectra for special plant during a certain period and under certain environmental conditions. The LF is of crucial scientific significance that should be established for plants grown with artificial light. More importantly, the LF is an important part of light environment management strategy (LEMS). The LEMS refers to a comprehensive management method of light environment, including light intensity, LF and photoperiod for one special plant, which should be established for plants with artificial light sources. It is assumed that the LF and LEMS will be preferentially applied in plant factory with entire artificial light source.

Keywords: Light formula (LF); necessary light spectrum; light-emitting diodes (LEDs); plant factory; light environment management strategy (LEMS)

INTRODUCTION

Protected vegetable cultivation has developed rapidly worldwide to meet the increasing demand for fresh vegetables. For example, it was estimated that the planting area of China of protected vegetables reached four million hectares in 2010 (Zhang, 2011). Nowadays, the protected facilities mainly include glasshouse, greenhouse, Chinese solar-greenhouse, plastic tents and plant factories for off-season vegetable cultivation. Plant factory is the top pattern of modern protected horticulture. Currently, dozens of plant factories, especially vegetable factories, are in operation in some countries, e.g. Japan, China, and Netherlands etc. As the dominant type, the vegetable factories trace their roots back to the Christensen Farm in Denmark, where production first started in 1957 (Takatsuji, 1986).

Light is a vital environmental factor that affects plant growth and development by acting on plants not only as the sole energy source of photosynthesis (acting on chlorophyll), but also as the kind of external signal (acting on cryptochrome, phototropin and the other photoreceptors) after being intercepted and absorbed by photosynthetic tissue. It is well known that plant growth and development are regulated by light quality, light intensity and photoperiod, so above three elements are key components of light condition. Under cover, besides photosynthetically active radiation (PAR) (400-700nm), the medium-wave ultraviolet light (UV-B, 280-320nm) long-wave ultraviolet light (UV-A, 320-400nm) are important spectral components for vegetable production. However, due to the absorption and obstruction effects of cover materials (glass and plastic film), only about 88% visible light and 15.9% to 21.1% ultraviolet irradiation is transmitted into protected systems, which resulted in significantly reduction in light intensity and even substantially modification in light spectrum composition (Nitz et al., 2004; Chen and Wu, 2008; Peng and Ai, 2010). In order to obtain an optimal light environment for vegetable growth, artificial light is needed to overcome the irregular diurnal light intensity changes of natural light in greenhouse or as the sole main light source. More importantly, UV-A, UV-B and UV-C irradiation levels in closed cultivation systems with artificial light, e.g. plant factory, are very low, since ultraviolet irradiation is not present in fluorescent lamps and light-emitting diodes (400-700nm). Therefore, detailed investigation on effects of light quality of PAR and ultravoilet irradiation on growth, physiological metabolisms and nutritional quality is necessary.

Generally, the improvement ofyield and nutritional quality of vegetables in protected facilities is the final target of light environment regulation. Based on current literatures, precise spatiotemporal management of light conditions (light quality, light intensity and photoperiod) can enhance yield and accumulation of health-beneficial phytochemicals in vegetables. Optimised light environment regulation can be environmentally friendly . High-quality vegetables can be produced in protected facilities. Liu et al.(2009) defined high-quality vegetables as vegetables not containing over-the-limit harmful substances (nitrate, nitrite, heavy metals and pesticides), and not having negative effects on human health after ingestion. Health-beneficial phyto-chemicals of vegetables are referred to some substances, such as ascorbic acid, carotenoids, phenolics, flavolds, which are influential beneficially to human health.

In greenhouses, the artificial light source is used to support the natural light, while the artificial light source itself entirely determines the light conditions in closed vegetable factories. So, the strategy for balancing the light requirements of vegetables to the light conditions is largely different in the two kinds of protected facilities. Comparatively, the latter is relatively easy to realize. In our view, it is possible to regulate the productivity and nutritional quality of horticultural crops in protected facilities. However, what kind of light quality or wavelength combinations is suitable for a specific horticultural crop? How to manage light quality, light intensity and photoperiod synergically? What is the light environment management strategy for various protected facilities? These issues is far fully investigated, understood and answered. Therefore, more work is urgently needed to clarify the relationship between light conditions (light quality, light intensity and photoperiod) and growth & quality at every developmental stages of plants in various protected systems. Apparently, closed plant factory with artificial lighting is the ideal protected facility as objective that should be studied on above issues preferentially, and study outcome is priorily applied in closed plant factory.

LIGHT FORMULA (LF): CONCEPT AND SIGNIFICANCE

As qualitative factor, light quality is the primary component in light condition. Based on current literatures, biologically active light quality can be classified into three zones located in spectrum, including PAR zone (400nm to 700nm), far-red zone (700nm to 750nm), and ultraviolet irradiation zone (200nm t0 400nm). PAR are comprised of red, blue, green, purple, yellow, orange, cyan, while ultraviolet irradiation is comprised of UV-A (320nm to 400nm), UV-B (280nm to 320nm), and UV-C (200nm to 280nm). Higher plants are empowered with an array of photoreceptors controlling diverse responses to light quality for plant growth and its photomorphogenesis. Generally, plants have three different photoreactions: (1) red and blue light used in photosynthesis; (2) blue light used in cryptochrome and phototropin reaction systems; (3) red and far-red light enabling reversible switching of the phytochrome system (Watanabe, 2011).

For natural light, light quality is often unsuitable for horticultural growth and health-beneficial phytochemical formation. Traditionally, both supplemental lighting in greenhouse and in closed vegetable factoris , usestable light conditions to maintain photosynthesis and photomorphogenesis, so light is a critical yet passive entity. With the development of light-emitting diode

(LED), the potential to actively implement dynamic lighting strategies to control plant growth and development, and nutritional quality holds great promise in the future of protected cultivation. LEDs can tailor illumination spectra according to plant requirements. Based on the advancements in LED illumination technology, here, we put forward light formula (LF) concept for guiding the optimization of light quality application in protected plant cultivation. LF is defined as an optimized light quality component aiming at high productivity or nutritional quality under protected production. Basically, LF of plants are subjected to species, cultivar, growth and developmental stages of plant, environmental conditions and manipulation target of yield & quality. In other words, LF is cultivar-specific, dynamic and adjustable accompanied with the corresponding biological processes of plants. Not only establishment of plant LF relies on LED technology development, also application of LF relies on LED lighting systems. Surely, LED, a sort of solid-state, narrow bandwidth lighting platforms, offer a unique opportunity to realize light quality precise managements according to LF. In the future, a set of specific LF should be studied and developed for specific horticultural crops, and LF may be adjusted throughout the entire life of plant to potentially optimize traits of interest, such as morphology, yield and nutritional quality.

LF and its management strategy are of great significance for closed plant factory production with artificial light. First, LF management will save energy by deleting extra light spectra; (2) LF management will maximize the yield and/or nutritional quality; (3) LF management is a basis for establish light environment management strategy (LEMS). Nowadays, LF and its management strategy is far perfectly established. Detailed studies on light fomula (LF) based on physiological requirement of plants one by one are urgently needed for high yield and quality targets.

Light-Emitting Diodes (LEDs): An Optimal Tool for LF Research

LF, in essence, is an optimized and integrated assembly of spectral component emitted from light sources that is suitable for plant productivity and nutritional quality formation. LED is an ideal tool to study light quality requirements of every horticultural crop. Their small size, durability, long operating lifetime, wavelength specificity, relatively cool emitting surfaces, and linear photon output with electrical input current make these solid-state light sources ideal for use in plant lighting designs. Because the output waveband of LEDs is much narrower than that of traditional sources of electric lighting used for plant growth, one challenge in designing an optimum light spectrum (Massa et al., 2008). LEDs have tremendous potential as supplemental or sole-source lighting systems for crop production. Several accepted advantages of LED can be classified. First, LED is energy-saving, small size and with high light efficiency; Second, LED is cold light source with low thermal radiation, so it can illuminate plants closely; Third, LED can emit multiple narrow-bandwidth monochromatic light with specific wavelengths, almost covering all biologically active light qualities. In addition, light quality biology and physiology have been gradually and extensively conducted worldwide by taking advantage of LED providing precise light spectrum and close illumination, which provide abundant data to establish LF. Totally, LED is a best candidate as light source for regulating plant biology, productivity and nutritional quality. Up to date, a number of studies and their findings are useful for selection of LED types and positioning for a variety of purposes depending on crop type and desired responses. To select specific wavelengths for a targeted plant response make LEDs more suitable for plant-based uses than many other light sources. More importantly, compound or mixed light quality according to LF can be realized through combination with various LEDs. Thus, using LED as artificial light source, light quality can be controlled with great precision as desired (Folta and Childers, 2008).

Light Formula (LF) for Protected Horticultural Plants: Primary Conclusion

LEDs have become an optimal tool for LF research of plants, particularly vegetables in closed plant factory or growth chamber since 1990s last century. It is well known that NASA's Kennedy Space Center and Purdue University had made great contributions for LED lighting technology (Folta and Childers, 2008; Massa et al., 2008). Today, scientists of many countries, e.g. Japan, USA and China, are positively working in this field. With rapid development of LED technology in 21th century, research on light biology and physiology of vegetables accelerated because more pure light wavelengths were emitted from novel LED lighting systems.

In 1990s, some initial studies focused on usefulness of LEDs as a sole source or as supplemental lighting for plant growth. For PAR, biological effects of red light plus blue light were extensively investigated on lettuce (Bula et al., 1991; Hoenmecke et al., 1992; Yanagi et al., 1996b; Yorio et al., 1998), strawberry (Yanagi et al., 1996a) and pepper (Brown et al., 1995), wheat (Goins et al., 1997) and rice (Matsuda et al., 2004). Goins et al. (1997) indicated that wheat can complete its life cycle under red LEDs alone, but larger plants and greater amounts of seed are produced in the presence of red LEDs supplemented with a quantity of blue light. Matsuda et al. (2004) also found that rice plants grown under a combination of red (660 nm) and blue (470 nm) LEDs sustained higher leaf photosynthetic rates than did leaves from plants grown under red LEDs only. Yorio et al. (1998) summarized previous blue light work and reported that yield of lettuce, spinach, and radish crops grown under red LEDs alone was reduced compared with when blue fluorescence was included to give the same final PPF. Schuerger et al. (1997) examined changes in leaf anatomy of pepper under different color combinations of light. They used red (660 nm) LEDs combined either with FR (735 nm) LEDs or BF lamps at the same PPF. Their results indicated that leaf thickness and number of chloroplasts per cell depended much more on the level of blue light than the red:FR ratio. Many horticultural and grain crops had been cultivated successfully under LED light sources with red light, blue light or red plus blue light quality, including lettuce, spinach, strawberry, tomato, cucumber, potato, wheat and rice as white light. Furthermore, mixed light quality comprised of suitable red light and blue light ratio did facilitate high productivity or nutritional quality, e.g. lettuce (Zhou et al., 2011), spinach (Ohashi-Kaneko et al., 2007), strawberry (Yanagi et al., 2006). Previous work has focused on the proportion of blue light required for normal plant growth as well as the optimum wavelength of red and the red/far-red ratio. Thus it can be concluded that, as white light, blue light and red light is necessary light quality for plant cultivation. Combination of red and blue lighting was an effective light source for several crops.

Other light qualities were also evidenced in yield or quality benefits in vegetable production under artificial lighting. Kim et al. (2004) found that the addition of 24% green light to red and blue LEDs enhanced lettuce growth since green light can better penetrate the plant canopy and potentially increase plant growth by increasing photosynthesis. The addition of green light could offer more benefits, since green light can better penetrate the plant canopy and potentially increase plant growth by increasing photosynthesis from the lower canopy leaves. Similarly, other light wavelengths , such as yellow, orange, purple, cyan and so on, may affect horticultural crops to some extent under certain conditions. This needs further investigations. Liu et al. (2011) found that cherry tomato seedlings were significantly stronger and shorter under ?, than under yellow, green and red were weaker and higher. Photosynthetic pigments were shown to have significant difference under respective light irradiations of LEDs. Taken together, red plus blue and red plus blue and green light of LEDs were shown to be beneficial factors for the growth and photosynthesis of cherry tomato seedlings.

Recently, usefulness of supplemental ultravoilet radiation on growth and nutritional quality of protected vegetables were also explored. Using UV-A, blue, green, red, and far-red LEDs, authors investigated the effects of different supplemental light qualities on phytochemicals and growth of baby leaf lettuce grown white fluorescent lamps as the main light source inside a growth chamber (Li and Kubota, 2008). Briefly, supplemental blue or UV-A could enhance accumulation of anthocyanins, supplemental blue also increased carotenoids concentration, supplemental R could increase phenolics concentration while supplemental far-red light could increase biomass, but result in lower phytochemical concentrations. A controlled light quality with an appropriate ratio of blue, red or far-red light quality provided as supplemental light may improve phytochemical content and biomass of plants grown under white light. Further studies are needed to describe the effects of different ratios of selected light qualities in both growth chamber and greenhouse conditions. The results demonstrated that supplemental light quality could be strategically used to enhance nutritional value and growth of baby leaf lettuce grown under white light (Li and Kubota, 2008). Via continuous lighting by LED, the contribution of red light to significant β-carotene expression and antioxidant activity for nutrition and health benefits are emphasized (Wu et al., 2007).

Also, the potential for increasing secondary compounds in vegetables by using supplemental selected UV irradiation have been extensively investigated (Tsormpatsidis et al., 2008; Voipio and Autio, 1995). UV radiation can be regarded as a stress factor which is capable of significantly affecting plant growth characteristics. Generally, plant height, leaf area, leaf length have been showed to decrease, whereas leaf thickness was increased in response to UV-B radiation (Rozema et al., 1997). Plants produce a wide range of flavonoids and related phenolic compounds which tend to accumulate in leaves of higher plants in response to UV radiation (Rozema et al., 1997). Plants may produce secondary products to protect them against UV light damage, but these metabolites also play an important role in human health. Phenolics, flavonoids and anthocyanins are responsible for antioxidant activity in fruits and vegetables (Cao et al., 1997).

Based on published literatures, red, blue, and compound white light are macro-necessary light spectrum, and purple, green, yellow and orange light are micro-beneficial light spectrum, while far-red light and ultravoilet light are beneficial light spectrum. Spectral changes of illumination evoked ifferent morphogenetic and photosynthetic responses, which can vary among different plant species. However, other light spectrum except photosynthetic active radiation, far-red and ultravoilet light is unvalid for plant cultivation. Our hypothesis is that one or two kind of necessary light quality, special micro-beneficial and beneficial light quality can make up a light formula for a kind of plant during certain period and certain environmental conditions. LF is a crucial scientific issue that should be established for protected cultivation plants with artificial light or plants that needed supplemental light.

LF: An Important Part of the Light Environment Management Strategy (LEMS)

Establishment of LF will lay a foundation for establishing light environment management strategy (LEMS) for given protected plant species cultivated with artificial light source. Besides LF, light intensity and photoperiod are key contents of the light environment management. The LEMS refers to a comprehensive management method of light environment, including LF, light intensity and photoperiod for entire life of one special plant species, which should be established to facilitate high productivity and good nutritional quality with artificial LED light sources. By using LED lighting systems, it is possible to continually adjust fluence rate, wavelength combinations, and photoperiods to actively manipulate plant morphology and production, rather than using a stable light condition traditionally.

The application of LEMS will be used in the plant factory for the advantages of entire artificial light source, cultivation with nutrient solution and intelligent environmental control. Furthermore, vegetable production, particularly leafy vegetable cutivation, will benefit preferentially from the use of LEMS in plant factory. Corresponding LED lighting systems for plant factory should be designed and developed. Ideally, for an optimum plant production and product quality, light environment have to be adapted to the needs of the plants at every moment controlled by LEMS.

Conclusion

With the development of LED illumination technology, light biology and physiology of protected horticultural crops were conducted extensively. Simultaneously, LEDs are bringing a breakthrough in artificial lighting in protected horticulture. LEDs provide a tool to facilitate the study and establishment of the LF and LEMS of every target horticultural crops. In return, LF will provide technique parameters for design of species-specific lighting regimes may help maximize plant productivity and benefits of food quality. Optimization of lighting system, establishement of LF and LEMS will drive plant cultivation in plant factory greatly either in productivity or nurtritional quality.

Acknowledgments

This study was supported by the National High Technology Research and Development Plan of China（863 Project, grant No. 2011AA03A114) and the Basic Scientific Research Fund of National Nonprofit Institutes (BSRF201004) and 2012-2013, Institute of Environment and Sustainable Development in Agriculture, CAAS.

References

Brown, C.S., A.C. Schuerger, and J.C. Sager. 1995. Growth and photomorphogenesis of pepper plants grown under red light-emitting diodes supplemented with blue or far-red illumination. *J. Amer. Soc. Hort. Sci.* 120:808–813.

Bula, R.J., R.C. Morrow, T.W. Tibbitts, D.J. Barta, R.W. Ignatius, and T.S. Martin. 1991. Lightemitting diodes as a radiation source for plants. *HortScience* 26:203–205.

Cao, G., Sofic, E., Prior, R.L., 1997. Antioxidant and prooxidaant behavior of flavonoids: structure–activity relationships. *Free Radic. Biol. Med.* 22, 749–760.

Chen L, Wu Z. 2008. Effects of UV-B on growth, yield and quality of pakchoi. *Journal of Plant Resources and Environment*.17：43-47.

Folta K.M., Childers K.S. 2008. Light as a growth regulator: controlling plant biology with narrow-bandwidth solid-state lighting systems. *HortScience*. 43: 1957-1963.

Goins, G.D., N.C. Yorio, M.M. Sanwo, and C.S. Brown. 1997. Photomorphogenesis, photosynthesis, and seed yield of wheat plants grown under red light-emitting diodes (LEDs) with and without supplemental blue lighting. *J. Expt. Bot.* 48:1407–1413.

Hemming S. 2011. Use of natural and artificial light in horticulture-interaction of plant and technology. *Acta Horticulturae* , 907: 25-35.

Kim H.H., Wheeler R.M., Sager J.C., Goins G.D. 2004. A comparison of growth and photosynthetic characteristics of lettuce grown under red and blue light-emitting diodes (leds) with and without supplemental green LEDs. *Acta Hort.* 659:467-475.

Li .J.C., Liu W.K., Yang Q.C. 2010. Strategic idea of replacing resources with environmental factors in agricultural production through protected agricultural technology. *Chinese Agricultural Science Bulletin* 26(3):283-285.

Liu W.K., Yang Q.C., Du L. F. 2009. Soilless cultivation for high-quality vegetables with biogas manure in China: feasibility and benefit analysis. *Renewable Agriculture and Food Systems*. 24(4):300–307.

Liu X.Y., Chang T.T., Guo S.R., Xu Z.G., Li J. 2011. Effect of different light quality of led on growth and photosynthetic character in cherry tomato seedling. *Acta Hort.*907: 325-330.

Liu XiaoYing, Guo ShiRong, Xu Zhiang , Jiao XueLei, Takafumi Tezuka. 2011Regulation of chloroplast ultrastructure, cross-section anatomy of leaves and morphology of stomata of cherry tomato by different light irradiations of LEDs. *Hortiscience*.45 (2):1-5.

Marcelis L.F.M., Snel J.F.H., de Visser P.H.B., et al. 2006. Quantification of the growth responses to light quantity of greenhouse grown crops. *Acta Horticulturae*711:97-104.

Massa G. D., Kim H.-H., Wheeler R. M., Mitchell C. A. 2008. Plant productivity in response to LED lighting. *HortScience*. 43(7): 1951–1 956.

Matsuda, R., K. Ohashi-Kaneko, K. Fujiwara, E. Goto, and K. Kurata. 2004. Photosynthetic characteristics of rice leaves grown under red light with or without supplemental blue light. *Plant & Cell Physiol.* 45:1870–1874.

Matsuda, R., K. Ohashi-Kaneko, K. Fujiwara, E. Goto, and K. Kurata. 2004. Photosynthetic characteristics of rice leaves grown under red light with or without supplemental blue light. *Plant & Cell Physiol.* 45:1870–1874.

Morrow, R.C. 2008. LED lighting in horticulture. HortScience 43:1947–1950.

Nitz, G. M., and Grubmuller, E., and Schnitzler,W. H. Differential flavoniod response to PAR and UV-B light in chive (*Allium schoenoprasum* L.). *Acta Hort.* , 2004, 659: 825–830.

Ohashi-Kaneko, K., Takase, M., Kon, N., Fujiwara, K., Kurata, K., 2007. Effect of light quality on growth and vegetable quality in leaf lettuce, spinach and komatsuna. *Environ. Control Biol.* 45, 189–198.

Peng Y, Ai X. A review on effects of UV-B increase on vegetables. *Modern Horticulture,* 2010, 6 : 16-17.

Rozema, J., Staaij, J.vd., Bjorn, L.O., Caldwell, M., 1997. UV-B as an environmental factor in plant life: stress and regulation. *Trends Ecol. Evol.* 12, 22–28.

Samuolienė G., Brazaitytė A., Urbonavičiūtė A., Šabajevienė G., Duchovskis P. The effect of red and blue ligjht component on the growth and development of frigo strawberries. *Zemdirbyste-Agriculture*, 2010,97(2): 99-104.

Schuerger, A.C., C.S. Brown, and E.C. Stryjewski. 1997. Anatomical features of pepper plants (Capsicum annuum L.) grown under red light-emitting diodes supplemented with blue or far-red light. *Ann. Bot.* (Lond.) 79:273–282.

Takatsuji M. *Vegetable plant factory*. Maruzen, Tokyo, 1986.

Tsormpatsidis, E., Henbest, R.G.C., Davis, F.J., Battey, N.H., Hadley, P., Wagstaffe, A., 2008. UV irradiance as a major influence on growth, development and secondary products of commercial importance in Lollo Rosso lettuce 'Revolution' grown under polyethylene films. *Environ. Exp. Bot.* 63, 232–239.

Wang Y.L., Wang X.L., Yu E.M. Effects of supplementary radiation of UV-B and red light on fruit quality of tomato in winter plastic greenhouse. *Acta Bot. Boreal.* -Occident. Sin, 2000, 20(4): 590-595.

Watanabe H. Light-controlled plant cultivation system in Japan-development of a vegetable factory using LEDs as a light source for plants. *Acta Horticulturae*, 2011, 907:37-44.

Wu M C, Hou C Y, Jiang C M, Wang Y T, Wang C Y,Chen H H,Chang H M. 2007.A novel approach of LED light radiation improves the antioxidant activity of pea seedlings. *Food Chemistry*, 101 (4) :1753-1758.

Yanagi, T., K. Okamoto, and S. Takita. 1996a. Effect of blue and red light intensity on photosynthetic rate of strawberry leaves. *Acta Hort.* 440:371–376.

Yanagi, T., K. Okamoto, and S. Takita. 1996b. Effect of blue, red, and blue/red lights of two different PPF levels on growth and morphogenesis of lettuce plants. *Acta Hort.* 440:117–122.

Yanagi, T., T. Yachi, N. Okuda, and K. Okamoto. Light quality of continuous illuminating at night to induce floral initiation of Fragaria chiloensis L. CHI-24-1. *Sci. Hort.*, 2006,109:309–314.

Yorio, N.C., G.D. Goins, H.R. Kagie, R.M. Wheeler, and J.C. Sager. 2001. Improving spinach, radish, and lettuce growth under red light-emitting diodes (LEDs) with blue light supplementation. *HortScience* 36:380–383.

Zhang Z.B. Development countermeasure of protected vegetables with low carbon production technology. In: Yang Q.C., Kozai T., Bot G.P.A. Protected Horticulture Advances and Innovations-Proceedings of 2011 the 2nd High-level International Forum on Protected Horticulture (Shouguang·China). *China Agricultural Science and Technology Press.* Beijing. 2011. 9-13.

In: Transgenic Plants
Editors: Shen Y. Zhu and Jiang L.Hu

ISBN 978-1-62257-245-8

Chapter 3

HARVEST-INDUCIBLE GENES AND PROMOTERS IN ALFALFA

Jian Zhang** *and Larry R. Erickson
Plant Agriculture Department, University of Guelph,
Guelph ON, Canada

ABSTRACT

The harvesting and storing of alfalfa is a routine practice in the agricultural industry worldwide. A study of the physiological and biochemical changes that occur after harvesting may help to understand how plants respond to this process.

To investigate gene expression in harvested alfalfa (*Medicago sativa*), cDNA from non-harvested and harvested plants in the field was subjected to subtractive hybridization to identify, in particular, those genes that are induced by the harvesting treatment. Three different genes, named, *hi7, hi11 and hi12,* were isolated and analysed. Northern blot analysis confirmed that *hi7, hi11* and *hi12* are strongly induced during a post harvest incubation period. The promoters of these genes were isolated and characterized. Reporter gene driven by the promoter has been transformed into transgenic plants. The hi12 promoter showed it is harvesting inducible.

Keywords: Harvest-inducible gene, promoter, transformation, Alfalfa

---

* Current address: Alberta Innovates-Technology Futures, Vegreville, AB, T9C1T4 Canada.

INTRODUCTION

Alfalfa (*Medicago sativa* L.) is probably native to Asia Minor and the Caucasus mountain area, and has been cultivated since antiquity. Alfalfa was likely cultivated before recorded history in a region spanning from China in the east to Spain in the west and from as far north as Sweden to as far south as North Africa. Recently, it has become an important crop in South Africa, Australia, New Zealand and North and South America [1]. Alfalfa is the most important forage legume crop in North America. In Ontario, forages were grown on about 2.5 million ha with a value of hundreds of millions of dollars of commercial hay. A major component of the forage crop in Ontario is alfalfa. Alfalfa is highly regarded for its feeding value which is due to its high protein content, and it is often called the "Queen of Forages" [2]. Alfalfa is harvested and stored primarily as hay or silage for use on farm. In all cases, the above ground part of the plant is cut near the base and laid in a swath in the field to wilt and reduce the water content, a process that can last for a period ranging from a few hours to a few days depending on weather conditions.

While lying in the swath, the harvested alfalfa plant tissue continues to function metabolically and the synthesis of mRNA and protein continues at least until the moisture level declines to ~20%.

Ferullo et al. have demonstrated that proteins are synthesized in harvested alfalfa plants, which are specific to harvesting conditions, and are not induced by wounding or water stress treatments on non-harvested plants [3]. However, the identity of these proteins was not determined and the biochemical and physiological aspects of protein metabolism after harvesting are largely unknown. Subtractive hybridization is a powerful method allowing one to obtain cDNA libraries enriched with transcripts that are present in one tissue or individual organism but not in another [4].

Straus and Ausubel were probably the first to report this method [5], and since then, subtractive hybridization has been found research applications in many disciplines ranging from identification of methylation changes during tumourigenesis [6] to isolation of diagnostic probes for infectious diseases [7].

In plants, it has been used to study the role of Chk1-a signaling element in the development of fungal pathogens on the plant host [8]. Matvienko et al. used a normalized, subtractive cDNA library enriched for transcripts differentially abundant in *Triphysaria versicolor* root tips treated with the allelopathic quinone 2,6-dimethoxybenzoquinone (DMBQ) to study the quinone detoxification mechanism in plants [9]. Zhang et al. constructed a cotton (Gossypium arboreum L.) drought-related cDNA library using using subtractive hybridization [10]. Salt-induced genes in *Suaeda maritima* were cloned using PCR-based suppression subtractive hybridization [11].

The expression of a foreign protein in a transgenic plant may be somewhat toxic or inhibitory to growth and development. The consequence of expressing such proteins with a constitutive promoter may range from regeneration of plants with a low concentration of that protein to distorted plant phenotype and sterility. A practical solution to this problem is to control expression of these proteins by inducible promoters, an approach well established in microbial systems.

An objective of the research conducted for this research was to investigate gene expression in harvested alfalfa tissue and specifically to isolate the genes that are induced by harvesting. Isolation and characterization of these genes and their regulatory elements could provide new information on the function of such genes in plants, as well as provide a novel system for expressing foreign proteins in harvested plant tissue. Since induction would occur only after harvesting, there would be no negative effect on plant growth and development due to the constitutive expression of foreign proteins.

MATERIALS AND METHODS

Plant Materials

Alfalfa plants, cv *Gala* (Northrup King, Canada) were harvested from a field in the second year of growth on a farm just south of Guelph, Ontario in sunny conditions and an air temperature around 25-28°C, just before noon. Plants were harvested from a random area in the field of approximately 1 m^2. The harvested whole plant samples were laid flat on the remaining stubble to wilt for approximately 1 h and reduce moisture content, as is done in normal harvest operations. The harvested samples taken from 10 plants were then covered with aluminium foil and stored at room temperature for 30 min, 2 h, 6 h and 24 h before RNA was extracted from leaves and petioles. The non-harvested control plants were immediately frozen in liquid nitrogen.

RNA Extraction

Total RNA was isolated from control samples and harvested samples using Trizol™ Reagent (Gibco BRL). The quality and quantity of total RNA were determined by formaldehyde-denatured agarose gel electrophoresis and spectroscopy. A QIAGene mRNA isolation kit was used to isolate poly A^+ mRNA from 1 ug total RNA from harvested tissue. The mRNA from non-harvested tissues was designated as the driver and the combined mRNA from tissue extracted 30 min and 6 h post-harvest was designated as the tester.

Subtractive Hybridization

A CLONTECH PCR-Select™ cDNA Subtraction Kit (CLONTECH Laboratories, Inc. Catalog # K1804) was used to generate a harvest-inducible subtractive cDNA library, according to the manufacturer's instructions. Tester and driver double stranded (ds) cDNAs were generated from both the harvest-induced mRNA population and the non-harvest induced mRNA population. Tester and driver ds cDNAs were then digested separately with restriction endonuclease *RsaI* to obtain short, blunt-ended molecules in the following typical reaction: 43.5 µl ds cDNA, 5.0 µl 10× *Rsa* I restriction buffer and 1.5 µl *Rsa* I (10 units/µl).

The reaction was mixed by a brief vortexing and centrifugation, followed by incubation at 37 °C for 2 h. The reactions were analyzed by agarose gel electrophoresis for efficiency of digestion and terminated by adding 2.5 µl of a 20 mM EDTA/glycogen mix and purified by a standard phenol/ chloroform/iosamyl alcohol procedure. The 50 µl reaction was divided equally into two 25 µl tester populations, each ligated with different adaptors, adaptor1 and adaptor 2r. However, no adaptor was ligated to the driver cDNA. Subtractive hybridization of the driver cDNA and adaptor 1 ligated tester cDNA and adaptor 2r ligated cDNA led to equalization and enrichment of differentially expressed sequences and generated two primary hybridization populations. The reaction was carried out as follows: 1.5 µl driver cDNA, 1.5 µl adaptor-1 ligated tester cDNA denatured at 98 °C for 1.5 min and incubated at 68°C for 8 h.

The second hybridization was conducted between these two populations without denaturing. The remaining equalized and subtracted ss tester cDNAs formed new hybridized molecules that were present only in the harvest-induced population. A PCR was used to obtain the desired harvest-inducible cDNA fragments. In the PCR reactions, primers specific to the adaptors were used and each reaction was comprised of 3 µl cDNA, 0.2mM dNTP mixture, 1.5 mM $MgCl_2$, 0.2 µM of each primer and 2 Units of *Taq* DNA polymerase (BRL Life Technologies). The PCR program was as follows: (1) 94 °C 3 min;

(2) 35 cycles of: 94 °C for 40 sec, 66 °C for 1 min, 72 °C for 1 min, (3) 72 °C for 10 min, and (4) stationary at 4 °C.

Cloning of Harvest-Inducible cDNA Fragments

The PCR products were cloned into TOPO™ TA cloning vector (InvitroGen, Canada) to generate the harvest-inducible cDNA library. The transformation was conducted according to the protocol provided with the TOPO™ TA Cloning kit®. The plates were incubated overnight at 37 °C and the white colonies were chosen for further analysis. Purified plasmid DNA, extracted by Genelute plasmid mini-pre kit (Sigma-Aldrich, St. Louis, Mo), was submitted for sequencing to Laboratory Services, a division of the University of Guelph. The PCR screening of bacteria, plasmid DNA purification, and sequencing procedures were carried out according to standard DNA manipulation methods [12].

Northern Blot Analysis

Northern blot analysis of harvest-inducible transcripts was conducted with probes excised from *hil1* and *hil2* cDNA clones using *Rsa* I. Total RNA was isolated from alfalfa plants following a post-harvest period of 0, 45 min, 2 h, 6 h and 24 h. Approximately 10 μg total RNA was electrophoresed in a formaldehyde denaturing 1 % agarose RNA gel. Blotting, hybrization and washing conditions were according to the manufacturer's instruction manual (GeneScreen, DuPont). A α-^{32}P isotope-labelled probe was hybridized with the membrane in 5X SSC, 10X Denhardt's solution, 0.1 % sodium pyrophosphate, 50 % formamide, 10 % dextran sulfate, 0.2 % SDS and 100 μg/mL denatured ss DNA at 42 °C for 24 h.

Isolation Promoter of the hil2 Gene by Genomic Walking

A CLONTECH Universal GenomeWalker kit (CLONTECH Laborotory Inc.) was used for isolating the upstream region of the hil2 gene according to the manufacturer's instructions. Five genomic walking libraries were generated from the genomic DNA digested with five restriction enzymes. The hil2 gene specific primer h12-98C (5'-CTTGTGACCAACTTCTTGATG

TCTTCC-3') was designed based on cDNA sequence data from Inoue et al (1998). The primary PCR reaction was conducted as follows: 7 cycles: 94°C for 25 s, 72°C for 3 min; 32 cycles: 94°C for 25 s, 67°C for 3 min and an additional final step of 67°C for 7 min. The secondary touchdown PCR with a nested primer provided by the kit was performed according to the manufacturer's instructions. The PCR was performed with the following cycle program: 5 cycles: 94°C for 25 s, 72°C for 3 min; 20 cycles: 94°C for 25 s, 67°C for 3 min and an additional final step of 67°C for 7 min. The PCR screening of bacteria, plasmid DNA purification, and sequencing procedures were carried out according to standard DNA manipulation methods.

a

```
                                   hi1 cDNA  fragment    551 BPS DNA
  1  ATCAAAAGGA AAATTTGGGT GGCAAGGGAT CTTCTCTTAT TATTTAAAAA AAGTACCTTG
 61  CCCGGGACGG CCCGCCCGGG CAGGTACAAA CTTTGATAAA CACTATGGAT GGATTTTACA
121  TTGCTCCTGC TTTTATGGAC AAGCTTGTTG TTCACATCAC CAAGAATTTC TTGACCCTAC
181  CCAACATCAA GGTTCCTCTC ATTCTTGGAG TCTGGGGAGG CAAAGGACAA GGAAAATCTT
241  TCCAAGCTGA GCTTGTCTTT GCTAAGATGG GAATCAACCC CATCATGATG AGTGCTGGAG
301  AGCTTGAAAG TGGAAATGCA GGAGAGCCAG CAAAATTGAT CAGACAGAGA TACCGTGAAG
361  CCTCTGACAT TATCAGGAAG GGAAAGATGT GTTGTCTGTT CATCAACGAT CTCGATGCAG
421  GAGCAGGTCG TATGGGTGGA ACCACACAAT ACACTGTCAA CAACCAGATG GTGAATGCCA
481  CACTCATGAA CATTGCTGAT AACCCCACCA ATGTCCAACT CCATGGTATG TACCTCGGCC
541  GCGACAGCGC T
```

b

```
                                   hi8 cDNA fragment    174 BPS DNA
  1  TCGAGCGGCC GCCCGGGCAG GTACTTTTTC CACATCCTGA ATCAGCTGCT AGACCAATCA
 61  CAATTGTCTG TGAGTCACCT GCTCCACATG TTATCAAGTA TCTTCTGTTG TTACTGTTAG
121  TTCTTCTGTT GTTGCTCTTT TTGGTTGTTG TGTACCTCGG CCGCGACCAC GCTA
```

c

```
                                   hi11 cDNA fragment    357 BPS DNA
  1  GGCTATTTGG GGTGGCAATG GCGCCGAGTT AGGACTATTC AAAATGATAT TAATGAGCTC
 61  AAGAAACAAC TCAAACCTGC ACCGGTCGCC CCTGATGGTA AAGCACTTGA AACTTCACCG
121  CCATCACCTG TTGAACTTCA AATCCAGAAA CTTACTGAGG AGAGGAAAGA GCTTATCAAA
181  GGTTCATACA GGGATAAACA CTTTAATGCT GGATCCATAC TTCTAGGATT TGGTGTCTTT
241  GAGGCTGTTG GTGTGAGGAC TCAACACATG GTTAAGGACA GGAAAGCTAT TTCCAGGTCC
301  ACATTTATTT GCAGGAGCAG GCATTACCGT CTTATGGGCA CTGGCAGCAG CTCTAGT
```

d

```
                                   hi12 cDNA fragment    566 BPS DNA
  1  TGTAATTAAA AAAGCTGGTG TTGATCACAA AATTGATTTC AGAGAAGGTC CAGCTCTTCC
 61  AGTTCTTGAT GAAATGATCA AAGACGAAAA GAATCATGGT AGCTACGATT TCATTTTTGT
121  GGATGCTGAC AAAGACAATT ACCTCAACTA CCATAAGAGG TTAATTGATC TTGTTAAAGT
181  GGGAGGTGTG ATCGGGTACG ACAACACCTT ATGGAATGGA TCTGTGGTTG CACCCCCTGA
241  TGCTCCATTG AGGAAGTATG TTAGGTACTA TAGAGATTTT GTTTTGGAGC TTAACAAGGC
301  TTTGGCTGTG GACCCTAGGA TTGAAATATG TATGCTTCCT GTTGGTGATG GAATCACTAT
361  CTGCCGTAGG ATCAAGTAAT TGGTTTGCAT GTGCACTATA TCATGTAATG CACTGCTCCA
421  CATTATTGAT CATTATTGTG TGGAAGCTAC AGAGCATTTA AAAGTCTTCA AGCCTTCTTG
481  TCTTTTGTTA TTTTTCTTCA ACATATTTGT GGTTGTAATT TTCTCTTGTC ATTGATATTG
541  AAACTTCGAA TAATTGAAAG TTATAT
```

Figure 1. The sequences of *hi1, hi8, hi11* and *hi12* cDNAs.

Results

Isolation of Harvest-Inducible cDNA Clones

A subtractive cDNA library was obtained from the hybridization of no-harvest and 6 h post-harvest induced mRNA populations (Figure 1). The library contained 12 colonies, from which four unique cDNA sequences were identified, *hi1, hi8, hi11* and *hi12* (*hi*-harvest-inducible. The *hi1* cDNA fragment of 551 bp in length, has high homology to *Arabidopsis* rubisco activase (*Rca1*) mRNA (AY130300) (Figure 2 a). Rubisco activase is a stromal protein that regulates the activation of rubisco, the most abundant enzyme in plants [13-14].

The *hi8* cDNA fragment was 174 bp in length and has high homology to the phosphoribulokinase precursor mRNA in common ice plant (*Mesembryanthemum crystallinum*), an enzyme which may play an important role in the regulation of the Calvin cycle in higher plants [15] (Figure 2 b).

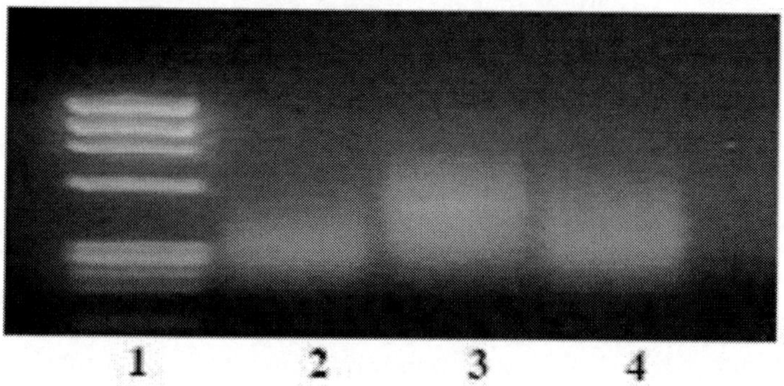

Figure 2. Secondary PCR amplification of harvest-inducible cDNA subtractive library. lane1: DNA ladder Φx174/Hind III; lane 2: PCR products amplified from subtractive hybridization between mRNA populations of non-harvested and harvested tissue at 30 min. lanes 3, 4: PCR products amplified from subtractive hybridization between mRNA populations from non-harvested and harvested tissue at 6 h.

The *hi11* cDNA was 357 bp in length and has high homology to a putative protein region with no known function in the genome database of *Arabidopsis thaliana* (Figure 2 c). The *hi12* cDNA was 348 bp long and was identical to the 3' end of the mRNA of *Medicago sativa* S-adenosyl-L-methionine: trans-caffeoyl-CoA 3-O-methyltransferase (CCOMT) mRNA [16-17], which catalyzes the formation of mono- or dimethoxylated lignin precursors (Figure 2 d).

The activity of this enzyme has been shown to increase from the first to the sixth internode in elongating stems of alfalfa and precedes the deposition of lignin.

Northern Blot Analysis

Northern analyses of *hi11* and *hi12* expression in harvested alfalfa tissue showed no detectable mRNA prior to harvest and a significant increase in transcript levels with an increase in the length of the post-harvesting incubation period. The harvesting treatment strongly induced *hi11* at 30 min, reaching a peak between 6 and 24 h after harvesting. The steady state level of *hi12* mRNA also showed a significant increase until 24 h post harvest, which was the longest post-harvest incubation period evaluated (Figure 3).

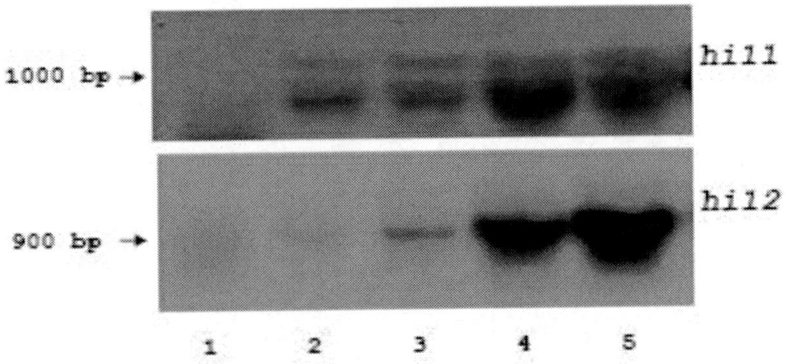

Figure 3. Northern blot analysis of *hi11* and *hi12* transcript levels in alfalfa leaf tissue following different post-harvest incubation periods. Arrows at left indicate expected size of mRNA. Approximately 10 μg RNA was loaded per lane. lane 1: 0 min; lane 2: 30 min; lane 3: 2 h; lane 4: 6 h; lane 5: 24h..

GUS Gene Construct with the hi12 Gene Promoter and Plant Transformation

A hi-12 *Bgl* II primer (5'-AGATCT-GATATGTGTTGAAGAGTTGA-3'), and hi12 *Sac* I primer (5'-GAGCTCGAATTCTTAAGTTACGTGA-3') were designed for amplification of the *hi12* promoter region. The GUS gene construct containing the *hi12* promoter was derived from pCAMBIA binary vectors. A 860 bp long *hi12* promoter fragment (5 to -854) was used for the *hi12*: GUS constructs [24]. The *hi12* promoter driven construct were mobilized into *Agrobacterium tumefaciens* strain C58, and tobacco, cv Pet H4, leaf discs were infected with bacterial cultures containing the individual constructs. The infected leaf discs were incubated at 28°C, in the dark for 4 days, and then transferred to antibiotic selection MS medium, containing 25 mg/l hygromycin or 100 mg/l kanamycin, at 24°C with a 16-h light/8-h dark cycle until shoots appeared.. The shoots were excised from callus tissue and maintained *in vitro* on one-half MS medium at 24°C with a 16-h light/8-h dark cycle, and then transplanted to the greenhouse. Histochemical staining for GUS activity was performed essentially as described by Jefferson et al. [23].

GUS Gene Activity Driven by hi12 Promoter

The *hi12* promoter (hi12-1301 construct) provided strong activity in transgenic tobacco and *Medicago truncatula* plants when the plant tissues had been subjected to a post harvest incubation period of 24 h (Figure 4).

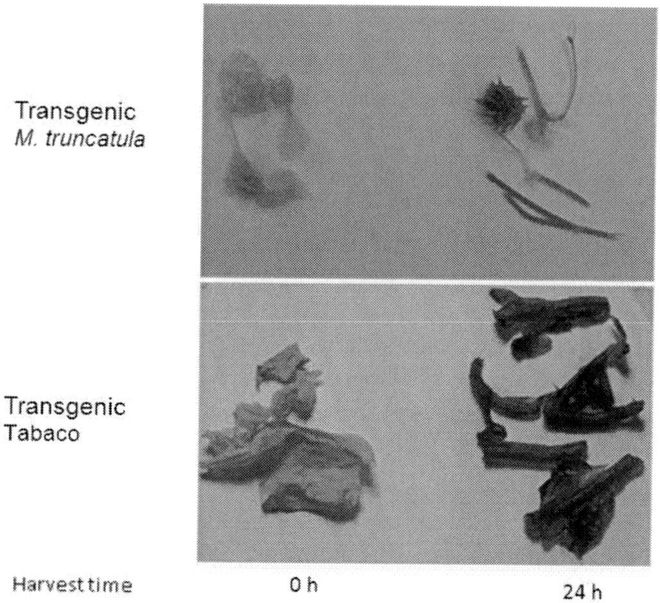

Figure 4. Histo-chemical staining for *β-glucuronidase* (GUS) activity in transgenic tobacco and *M. truncatula* leaf tissue, which containing the *hi12* promoter-driven construct following a post-harvest incubation period of 0h-0 hour and 24h-24 hours.

Discussion

An underlying hypothesis of this research was that the expression of some genes is up-regulated in harvested alfalfa tissue, and an objective was to isolate and characterize some of these genes. cDNA clones of harvest-inducible genes in alfalfa were isolated using PCR-based subtractive hybridization, a technique used to identify genes that are expressed in one mRNA population but not another [4]. In this process, all cDNAs that were not induced by harvesting were eliminated by hybridization. The subtracted cDNA library obtained by hybridization of cDNA populations generated from non-harvested and harvested alfalfa tissue contained several genes that were induced by the harvesting treatment. This was demonstrated by the Northern blot analysis, in which transcript levels of the genes showed a significant increase during a 6-24 h post-harvest incubation period compared to low levels in non-harvested alfalfa tissue. These results support the hypothesis that there are harvest-inducible or harvest-associated genes in the alfalfa plant. There were 12 clones generated from subtractive hybridization; this relatively small number of unique clones could be due the low number of genes involved in the harvesting response in alfalfa. The efficiency and specificity of the hybridization and cloning procedures, however, were verified in a parallel control experiment provided by the Clonetech kit with human skeletal mRNA, where a single clone was isolated. Moreover, the small number of clones is consistent with the small number of proteins (11) that Ferullo et al. found to be specific to the harvesting response [3].

The harvest-inducible cDNA clones, which were identified in this study, showed a broad range of homologies, and could be divided into three classes: response to light, general response to stress and others. Based on their homologies, *hi1*, and *hi8* are possibly involved in light response. *hi1* is identical to the alfalfa rubisco activase (RCA, AAN15946), but not yet fully characterized. Rubisco activase is an ATP-dependent enzyme, which appears to play a key role in photosynthesis [18-19]. Consequently, the enzyme is abundant in photosynthetic tissue, and probably accounts for 2 % of total soluble protein from spinach leaves [20]. Although it is evident that *hi1* responds to the harvesting treatment, the function of this enzyme in such tissue has yet to be determined. It has been suggested, however, that RCA is involved in the activation of rubisco through carbamylation. Conceivably increased RCA availability in harvested tissue may reflect a general metabolic shift from photosynthesis to a more catabolic process of remobilization of amino acids regenerated from the photosynthetic apparatus. Based on homologies *hi12* would appear to be associated with one or more general stress response functions in plants. *hi12* is the gene for the S-adensoyl-L-methionine-trans-caffeyol-CoA-O- methytransferase (CCOMAT) in alfalfa, an enzyme known to be involved in lignin synthesis in stems and implicated in mounting a physical defence to pathogen infection [21-22]. *hi11* cannot yet be associated with any known function on the basis of homology.

The use of harvest-inducible promoters in alfalfa may provide a unique and powerful inducible expression system in transgenic plants. Such a system would provide a measure of biological containment, as the foreign proteins are not produced in the transgenic plant until the application of the harvesting treatment, at which time the plant tissue is collected and stored. In addition, the crop can be harvested and stored by using conventional farm equipment and facilities. However, a much more complete analysis of transgenic plants containing foreign genes controlled by *hi* promoters is needed to verify the utility of such s system. The basal level of expression of reporter gene, such as GUS, needs to be determined in a range of tissue and under a range of conditions, such as drought, heat, cold and wounding by insects or pathogen.

ACKNOWLEDGMENTS

Extensive support for this research was provided by the Natural Sciences and Engineering Research Council of Canada, Strategic Grants Program.

REFERENCES

[1] D.A. Samac, S. Austin-Phillips, Alfalfa (Medicago sativa L.). *Methods Mol. Biol.* 343 (2006) 301-311.

[2] D.K. Barnes, B.P. Goplen, J.E. Baylor, "Highlights in the USA and Canada" in *Alfalfa and Alfalfa Improvement.* eds, American Society of Agronomy, Inc. Madison, 1988, pp1-22.

[3] J. Ferullo, L. Verzina, Y. Castonguay, G. Allard, P. Madeau, C. Willemot, S. Laberge, Post-harvest alteration of *in vitro* translatable mRNA population in alfalfa (*Medicago. sativa* L.). *Crop Sci.* 36 (1996) 1011-1016.

[4] B. Blumberg, J.C. Belmonte, Subtractive hybridization and construction of cDNA libraries. *Methods Mol. Biol.* 461 (2008) 569-587.

[5] D. Straus, F.M. Ausubel, Genomic subtraction for cloning DNA corresponding to deletion mutations. *Proc. Natl. Acad. Sci. USA* 87 (1990) 1889-1893.

[6] G.J. Brock, T.H. Huang, C.M. Chen, K.J. Johnson, A novel technique for the identification of CpG islands exhibiting altered methylation patterns. *Nucleic. Acids Res.* 29 (2001) e123.

[7] Y.N. Parsons, S. Panagea, C.M. Smart, M.J. Walshaw, A. Hart, C. Winstanley, Use of subtractive hybridization to identify a diagnostic probe for a cystic fibrosis epidemic strain of Pseudomonas aeruginosa. *J. Clin. Microbiol.* 40 (2002) 4607-4611.

[8] S. Lev, B.A. Horwitz, A miotogen-activated protein kinase pathway modulates the expression of two cellulose genes in Cochliobolus heterostrophus during plant infection. *Plant Cell* 15 (2003) 835-844.

[9] M. Matvienko, M.J. Torres, J.I. Yoder, Transcriptional responses in the hemiparasitic plant *Triphysaria versicolor* to host plant signals. *Plant Physiol.* 127 (2001) 272-282.

[10] L. Zhang, F.G. Li, C.L. Liu, C.J. Zhang, X.Y. Zhang, Construction and analysis of cotton (*Gossypium arboreum* L.) drought-related cDNA library. *BMC Res. Notes* 2 (2009) 120.

[11] B.B. Sahu, B.P. Shaw, Isolation, identification and expression analysis of salt-induced genes in Suaeda maritima, a natural halophyte, using PCR-based suppression subtractive hybridization. *BMC Plant Biol.* 9 (2009) 69.

[12] J. Sambrook, D.W. Russell, Molecular Cloning; A laboratory Manual, Cold Spring Harbor laboratory Press, New York, 2001.

[13] N. Zhang, R.P. Kallis, R.G. Ewy, A.R. Portis Jr., From the cover: light modulation of Rubisco in Arabidopsis requires a capacity for redox regulation of the larger rubisco activase isoform. *Proc. Natl. Acad. Sci. USA* 99 (2002) 3330-3304.

[14] R.F. Sage, D.A. Way, D.S. Kubien, Rubisco, Rubisco activase, and global climate change. *J. Exp. Bot.* 59 (2008) 1581-1595.

[15] N. Wedel, J. Soll, B.K. Paap, CP12 provides a new mode of light regulation of Calvin cycle activity in higher plants. *Proc. Natl. Acad. Sci. USA* 94 (1997) 10479-10484.

[16] K. Inoue, V.J. Sewalt, G.B. Murray, W. Ni, C. Sturzer, R.A. Dixon, Developmental expression and substrate specificities of alfalfa caffeic acid 3-O-methyltransferase and caffeoyl coenzyme A 3-O-methyltransferase in relation to lignification. *Plant Physiol.* 117 (1998) 761-770.

[17] Q. Yang, H.X. Trinh, S. Imai, A. Ishihara, L. Zhang, H. Nakayashiki, Y. Tosa, S. Mayama, Analysis of the involvement of hydroxyanthranilate hydroxycinnamoyltransferase and caffeoyl-CoA 3-O-methyltransferase in phytoalexin biosynthesis in oat. *Mol. Plant Microbe. Interact.* 17 (2004) 81-89.

[18] M.E. Salvucci, W.L. Ogren, The mechanism of rubisco activase: insights from studies of properties and structure of the enzyme. *Photosynth. Res.* 47 (1996) 1-11.

[19] S. Eberhard, G. Finazzi, F.A. Wollman, The dynamics of photosynthesis. *Annu. Rev. Genet.* 42(2008) 463-515.

[20] S.P. Robinson, V.J. Streusand, J.M. Chatfield, A.R. Portis Jr., Purification and assay of rubisco activase from leaves. *Plant Physiol.* 88 (1988) 1008-1014.

[21] R.A. Dixon, N.L. Paiva, Stress-induced phenylpropanoid metabolism. *Plant Cell* 7 (1995) 1085–1097.

[22] J.L. Ferrer, M.B. Austin, C. Stewart Jr., J.P. Noel, Structure and function of enzymes involved in the biosynthesis of phenylpropanoids. *Plant Physiol. Biochem.* 46 (2008) 356-370.

[23] RA Jefferson, TA Kavanagh, MW Bevan, GUS fusions: beta-glucuronidase as a sensitive and versatile gene fusion marker in higher plants. *EMBO J.* (1987) 6:3901-3907.

[24] J. Zhang, L.R. Erickson, Harvest-inducibility of the promoter of alfalfaS-adenosyl-Lmethionine: trans-caffeoyl-CoA3-O-methyltrans-ferase gene. *Mol. Biol. Rep.* (2012) 39:2489–2495.

In: Transgenic Plants
Editors: Shen Y. Zhu and Jiang L.Hu

ISBN 978-1-62257-245-8

Chapter 4

TOXIC IMPACTS OF THREE VETERINARY ANTIBIOTICS ON SEED GERMINATION AND GROWTH AS WELL AS NUTRITIONAL QUALITY OF VEGETABLES

Lian Feng Du[1] and·Wen Ke Liu[2,*]

[1]Institute of Plant Nutrition and Resources, Beijing Academy of Agriculture and Forestry Sciences, Beijing, China

[2]Environment and Sustainable Development in Agriculture, Chinese Academy of Agricultural Sciences, Key Lab for Agro-Environment and Climate Change, Ministry of Agriculture, Beijing, China

ABSTRACT

Two experiments were conducted to investigate the impacts of three antibiotics on seed germination, hydroponic growth and nutritional qualities of vegetables. Seed germination test of three vegetables (cucumber, rape and Chinese cabbage) with six concentrations of tetracycline (TC), chlorotetracycline (CTC) and oxytetracycline (OTC) in form of hydrochlorides was carried out to study the impacts of these antibiotics on root elongation. Hydroponic experiment was designed to explore the growth inhibition and nutritional quality responses of lettuce

---

* Corresponding author: liuwke@163.com.

grown in greenhouse with presence of various concentrations of CTC and OTC. The results showed that three antibiotics inhibited root elongation of germinated seeds, and the inhibitory rates depended on the drug species, antibiotic concentrations and vegetable species. Shoot and root biomass of lettuce were significantly reduced when antibiotics added in the nutrient solution. In addition, the soluble sugar content in leaves and petiole, and soluble protein in petiole were modified by antibiotic pollution. To conclude, antibiotic pollution of soil and water will adversely impact the seed germination and growth of protected vegetables grown in soil or hydroponics.

Keywords: Antibiotics, seed germination, vegetable ecotoxicity

INTRODUCTION

Globally, besides as human medicines, an increasing amount of antibiotics as veterinary drugs and feed additives were used annually for disease prevention & treatment and growth promotion in many countries with the rapid development of breeding industry (livestock breeding and aquaculture) (Cromwell et al. 1996; Sarmah et al. 2006; Cabello et al. 2006; Aust et al. 2008). As a result, mostly ingested antibiotics (40-90%) and their metabolites, as emerging persistent contaminants, were excreted with urine and feces, and subsequently disseminated into environmental compartments. Up to date, antibiotic contamination of effluents of sewage treatment plants (Giger et al. 2003; Göbel et al. 2004), surface water (Christian et al. 2003; Kolpin et al. 2002) , manures (Hu et al. 2008; Arikan et al. 2009), soils (Hamscher et al., 2002; Schlüsener et al., 2003; Thiele-Bruhn and Aust 2004) and groundwater (Hamscher et al., 2005) were extensively reported. More importantly, significant amount antibiotics and their bioactive metabolites or degradation products were introduced in agro-ecosystems through antibiotics-polluted animal manures and irrigation water. Nowadays, a large quantity of organic manures were heavily used in protected vegetable production systems (PVPS, e.g. greenhouses and plastic-covered tunnels) and organic vegetable production systems (OVPS) for vegetable production among agro-ecosystems in some countries, e.g. China (Shi et al. 2010; Hu et al. 2010). Therefore, high-level antibiotics accumulated in soils of PVPS and OVPS as revealed by Hu et al. (2010). As a results, accumulation and transport of antibiotics in soil-vegetable systems posssibly poses great risks on crop fitness, soil ecosystems, and quality of plant-based products. Hu et al. (2010) firstly investigated the occurrence and seasonality of antibiotics from manure, soil, vegetables and groundwater, and explored the migration of antibiotics from manure to soil, from soil to vegetables and groundwater. Shi et al. (2010) and Hu et al. (2010) firstly put forward the antibiotic pollution issues stemming from organic manures in protected and organic soil-vegetable systems in China. So, ecotoxic impacts of antibiotic contaminants in PVPS and OVPS need to be studied intensively for ensuring the sustainable production.

Recently, some results showed that antibiotics impacted seed germination and growth of some crops, and soil enzymatic activities (Jin et al., 2009; Liu et al., 2009). However, there is limited information about the ecotoxic impacts of antibiotics on protected vegetables, even less on hydroponic vegetables. In this study, two experiments were conducted to investigate the impacts of three antibiotics on seed germination, and hydroponic lettuce growth and quality parameters. In the first experiment, seed germination test of three vegetables (cucumber, rape and Chinese cabbage) with six concentrations of tetracycline (TC), chlorotetracycline (CTC) and oxytetracycline (OTC) in form of hydrochlorides was carried out to study the impacts of these antibiotics on seed germination. In the second experiment, hydroponic experiment was designed to explore the growth inhibition and nutritional quality responses of lettuce grown in greenhouse with presence of various concentrations of CTC and OTC.

Materials and Methods

Materials

All the test three veterinary pharmaceuticals, including tetracycline (TC), chlorotetracycline (CTC) and oxytetracycline (OTC) in form of hydrochloride, were purchased from Sigma-Aldrich Chemical Co. Four vegetable species were used in two experiments, including cucumber, Chinese cabbage, lettuce and rape, and the cultivars used are Jingyu1, Jinlü80, Italian cultivar and Beiji, respectively.

Experiment 1

Impacts of three antibiotics (tetracycline hydrochloride, chlorotetracycline hydrochloride and oxytetracycline hydrochloride) on germination of cucumber (Jingyu1), Chinese cabbage (Jinlü80) and rape (Beiji) were investigated in experiment 1. In the experiment, six concentrations of antibiotics were designed, including 0, 10, 50, 100, 250, 500 mg/L, respectively. Fifty-four treat-ments were included, and each treatment replicated three times.

Eight seeds for cucumber and ten seeds for Chinese cabbage and rape were disinfected by 10% H_2O_2 for ten minutes. After rinsing fully by tap water, seven seeds for cucumber and ten seeds for Chinese cabbage and rape were placed on a filter paper which was moistened with drug solutions and set down in a Petri dish (diameter 9cm). All the Petri dishes were cultured in a dark culture chamber. The temperature of the dark culture chamber were controlled within 30±0.5℃. The fourth day after cultivation, the root length for each replicate was recorded and averaged, then root length of three replicates of each drug concentration treatment was averaged again. Using the averages of all treatments, inhibitory rates were calculated as followed:

The inhibitory rate (IR) of the three vegetables was calculated as followed: Inhibitory rate (IR) = (Root elongation in control treatment-Root elongation in drug treatment)/ Root elongation in control treatment.

Experiment 2

Impacts of CTC and OTC on growth and quality parameters of hydroponic lettuce were investigated in experiment 2. Five treatments were designed, including 0, 2.5 and 5.0 mg/L CTC (CK, CTC-2.5 and CTC-5), and 2.5, 5.0 mg/L OTC (CTC-2.5 and CTC-5), respectively. All antibiotics were fully dissolved in water and then mixed into nutrition solution. Each treatment replicated three times. The hydroponic nutrient solutions were prepared with distilled water and analytical reagents. The nutrient solution is composed of 0.75 K_2SO_4, 0.5 KH_2PO_4, 0.65 $MgSO_4$, 0.1 KCl, 5.0 Ca $(NO_3)_2$, 1.0×10^{-3} H_3BO_3, 1.0×10^{-3} $MnSO_4$,1.0×10^{-4} $CuSO_4$, 5.0×10^{-6} $(NH_4)_6Mo_7O_{24}$, 1.0×10^{-3} $ZnSO_4$, 0.1 EDTA-Fe (mmol/L). The experiment was carried out in a glasshouse, with a temperature ranging from 15 to 30℃.

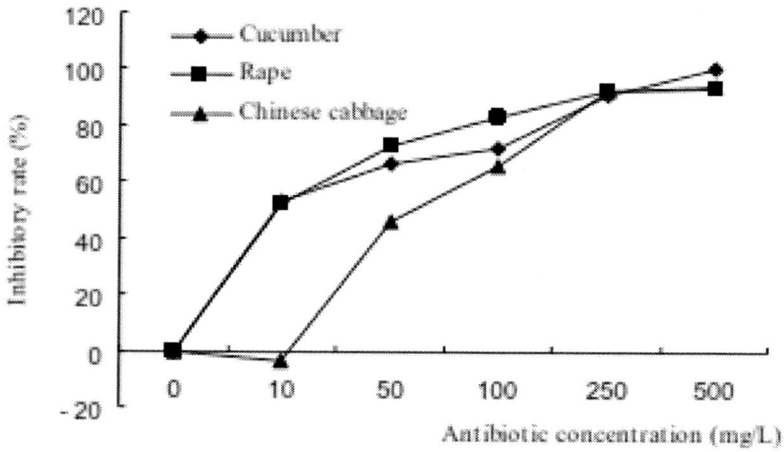

Figure 1. Inhibitory rates of tetracycline on root elongation of three vegetable seeds.

Lettuce (*Lactuca sativa*) , an Italian cultivar, was transplanted and cultivated in the experiment pot (32cm×24cm×11cm) supplied with 6L nutrient solution after 20 days cultivation in a seedling tray. Two lines and six holes were made on the lid at an equably distance, and a central one was used as a gasvent to aerate the nutrient solution by a air-pump (Atman EP-9000) to ensure the supply of soluble oxygen around the roots. Five lettuces in one hydroponic pot were planted. The experiment started on Sept. 13, 2010. The lettuces were harvest after thirty days. Lettuce shoot was separated three parts, i.e. new leaf, expanded leaf and petiole for nitrate concentration determination. Also the SPAD of new and expanded leaf were determined with Minolta SPAD-502 chlorophyll meter (made in Japan). The soluble sugar content and soluble protein content in leaf and petiole were measured by the methods described in Li et al. (2000). Also the SPAD of new and expanded leaf were determined with Minolta SPAD-502 chlorophyll meter (made in Japan).

RESULTS

Impacts of Antibiotics on Root Elongation of Three Vegetables

Impacts of TC on Root Elongation of Three Vegetables

The TC inhibited the root elongation of cucumber, rape and Chinese cabbage, and the inhibitory rates grew with the increase of the drug concentration (Figure 1). The tolerance of Chinese cabbage seed to three antibiotics is higher than cucumber and rape, particularly at the low drug concentration (<250mg/L). Almost all vegetables lost germination vitality when the drug concentration climbed to 500mg/L.

Impacts of OTC on Root Elongation of Three Vegetables

The CTC inhibited the root elongation of cucumber, rape and Chinese cabbage, and the inhibitory rates grew with the increase of the drug concentration (Figure 2). The tolerance of cucumber seed to three antibiotics is higher than Chinese cabbage and rape, particularly at the low drug concentration (<250mg/L). Almost all vegetables lost germination vitality when the drug concentration climbed to 500mg/L.

Impacts of CTC on Root Elongation of Three Vegetables

The CTC inhibited the root elongation of cucumber, rape and Chinese cabbage, and the inhibitory rates grew with the increase of the drug concentration (Figure 3). The tolerance of Chinese cabbage seed to three antibiotics is higher than cucumber and rape at the low drug concentration (<250mg/L). However, cucumber presented higher tolerance than Chinese cabbage and rape at high drug concentration (>250mg/L). In particular, at concentration 500mg/L of CTC, root elongation of cucumber was going on, but other vegetable species had ceased.

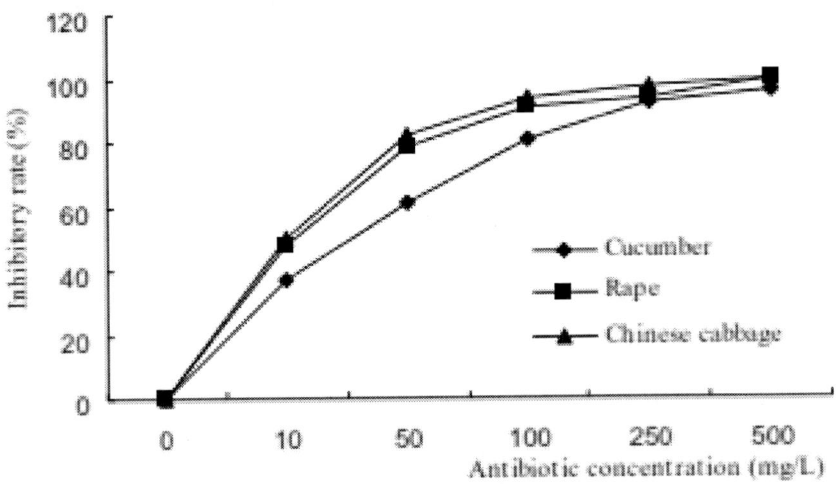

Figure 2. Inhibitory rates of oxytetracycline on root elongation of three vegetables.

Impacts of Antibiotics on Growth and Quality Indices of Hydroponic Lettuce

Impacts of CTC and OTC on Hydroponic Growth and Leafy SPAD of Lettuce

The results of hydroponic experiment showed that two antibiotics inhibited the shoot and root growth, and leafy SPAD of lettuce compared to the control (Table 1). Shoot biomass of lettuce declined more when OTC was presented than CTC. There are no significant concentration effects for two drugs in growth inhibition.

Impacts of CTC and OTC on Soluble Sugar Content of Lettuce

Soluble sugar content of leaf baldes and petioles of lettuces were influenced by antibiotics (Table 2). CTC-2.5 and CTC-5.0 treatments improved the soluble sugar content in leaf baldes significantly compared with the control. However, the soluble sugar content of leaf blades of treatment OTC-5.0 was slightly decreased. In addition, CTC-5.0, OTC-2.5 and OTC-5.0 treatments markedly increased the soluble sugar content in petioles of lettuces.

Table 1. Impacts of CTC and OTC on leafy SPAD and biomass of lettuce

Treatments	Leafy SPAD	Shoot biomass (FW)	Root biomass (FW)
CK	15.2a	28.74a	4.30a
CTC-2.5	13.0a	26.40ab	3.19b
CTC-5.0	13.8a	27.01ab	3.35ab
OTC-2.5	14.2a	20.77b	3.31ab
OTC-5.0	14.1a	19.54b	3.10b

Table 2. Impacts of CTC and OTC on nitrate, soluble sugar and soluble protein contents of lettuce

Treatments	Soluble sugar content (%, FW)		Soluble protein content (mg/kg, FW)	
	Leaf blades	Petioles	Leaves	Petioles
CK	0.17bc	0.15b	3.63a	0.64b
CTC-2.5	0.30a	0.18b	3.54a	0.97a
CTC-5.0	0.25ab	0.61a	3.65a	0.84ab
OTC-2.5	0.17bc	0.46ab	3.61a	1.03a
OTC-5.0	0.15c	0.62a	3.64a	1.01a

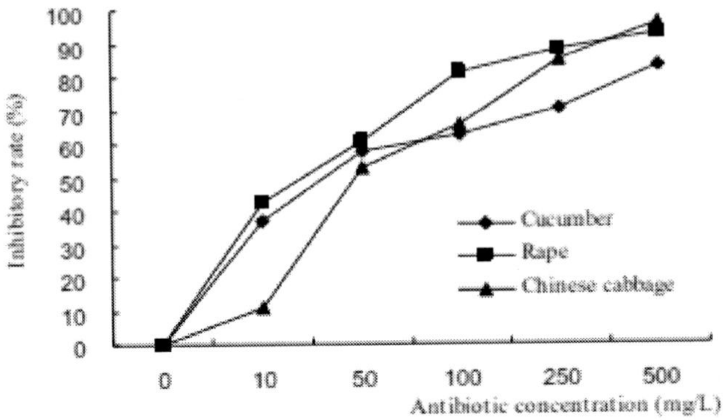

Figure 3. Inhibitory ratesof chlortetracycline on root elongation of three vegetables.

Impacts of CTC and OTC on Soluble Protein Content of Lettuce

Soluble protein content of leaf baldes of lettuces were not influenced by antibiotics, but that in petioles was altered (Table 2). All antibiotic treatments slightly or significantly increased the soluble protein content in petioles significantly compared with the control.

DISCUSSION

Antibiotics have just aroused attention as potential pollutants in past two decades (Halling-Sørensen et al. 1998; Kümmerer 2001), and very little information has been known about the soil exposure quantity, soil fates and environmental problems including ecological risks and human health damage in agro-ecosystems, particularly organic agriculture and protected agriculture. Based on the facts that PVPS and OVPS are extensively contaminated by antibiotics caused by fertilization organic manures and irrigation, ecotoxicity of antibiotic contaminants in PVPS and OVPS should be paid much attention to and be intensively studied. Present data showed that antibiotics could inhibited seed germination of vegetables, which indicates that accumulated antibiotics in PVPS and OVPS soil may impact seed germination of the sown seeds. In addition, our results also found that antibiotic-contaminated water deterious affected the growth of hydroponic lettuce, which suggests that water contamination by antibiotics may influence the productivity of vegetable productivity in hydroponics. As previous reported, antibiotics could inhibted seed germination and crop growth, such as wheat, Chinese cabbage, tomato, sweet oat, rice and cucumber (Jin et al. 2009; Liu et al. 2009). So far, only few antibiotic species were examined for their ecotoxicity to limited crop species, and the phytotoxic effects of antbiotics varied between the antibiotics and between plant species (Jin et al. 2009; Liu et al. 2009). Therefore, more investigations should be done in the future to fully understand the ecotoxicity of antibiotics in agro-ecosystems, particularly in PVPS and OVPS. Besides biomass inhibtion, we first revealed that antibiotics affected suluble sugar content and soluble protein content in leaves of hydroponic lettuce. That means that antibiotics may impact hydroponic growth through interfering in physiological processes, resultant nutritional quality changes. From present results, we suppose that antibiotic pollution of soil and water will adversely impact the seed germination and growth of protected vegetables grown in soil or hydroponics.

ACKNOWLEDGMENTS

This work was supported by the Project of Research & Development Program for China's Sustainable Agriculture (II), 2009-2014. Also the author want to thank the intern students of Capital Normal University, Yuan Gao and Wen Pan for their help in sample analysis.

REFERENCES

Arikan O., Mulbry W., Ingram D., Millner P. (2009) Minimally managed composting of beef manure at the pilot scale: effect of manure pile construction on pile temperature profiles and on the fate of oxytetracycline and chlortetracycline. *Bioresource Technology* 100:4447–4453.

Aust M. O., Godlinski F., Travis G. R., Hao X., McAllister T. A., Leinweber P. (2008) Distribution of sulfamethazine, chlortetracycline and TYL in manure and soil of Canadian feedlots after subtherapeutic use in cattle. *Environmental Pollution* 156:1243–1251.

Cabello F. C. (2006) Heavy use of prophylactic antibiotics in aquaculture: a growing problem for human and animal health and for the environment. *Environ. Microbiol.* 8:1137–1144.

Christian T., Schneider R. J., Farber H. A., Skutlarek D., Meyer M. T., Goldbach H. E. (2003) Determination of antibiotic residues in manure, soil, and surface waters. *Acta Hydrochim. Hydrobiol.* 31:36–44.

Cromwell G. L., Davis G., Morrow W., Primo W. E. M., Rozeboom R. A., Sims D. W., Stanisiewski M. D., Ho E. P. (1996) Efficacy of the antimicrobial compound U-82,127 as a growth promoter for growing-finishing pigs. *J. Anim. Sci.* 74 (6):1284–1287.

Giger W., Alder A. C., Golet E. M., Kohler H. P. E., McArdell C. S., Molnar E., Siegrist H., Suter M. J. F. (2003) Occurrence and fate of antibiotics as trace contaminants in wastewaters sewage sludge and surface waters. *Chimia* 57:485-491.

Göbel A., McArdell C. S., Suter M. J. F., Giger W. (2004) Trace determination of macrolide and sulfonamide antimicrobials a human sulfonamide metabolite and trimetoprim in wastewater using liquid chromatography coupled to electrospray tandem mass spectrometry. *Anal Chem* 76:4756-4764.

Hamscher G., Pawelzick H. T., Hoper H., Nau H. (2005) Different behavior of tetracyclines and sulfonamides in sandy soils after repeated fertilization with liquid manure. *Environ Toxicol Chem* 24:861–868.

Halling-Sørensen B., Nors Nielsen S., Lanzky P. F., Ingerslev F., Holten Lützhøft H. C., Jørgensen S. E. (1998) Occurrence, fate and effects of pharmaceutical substances in the environment - *A review. Chemosphere* 36:357-393.

Hu X. G., Zhou Q. X., Luo Y. (2010) Occurrence and source analysis of typical veterinary antibiotics in manure, soil, vegetables and groundwater from organic vegetable bases, northern China. *Environmental Pollution* 158(9):2992-2998.

Hu X. G., Zhou Q. X., Xu L. (2008) Determination of thirteen antibiotics residues in manure by solid phase extraction and high performance liquid chromatography. *Chinese Journal of Analytical Chemistry* 36(9): 1162-116.

Jin C. X., Chen Q. Y., Sun R. L., Zhou Q. X., Liu J. J. (2009) Eco-toxic effects of sulfadiazine sodium, sulfamonomethoxine sodium and enrofloxacin on wheat, Chinese cabbage and tomato. *Ecotoxicology* 18:878–885.

Kolpin D. W., Furlong E. T., Meyer M. T., Thurman E. M., Zaugg S. D., Barber L. B. (2002) Pharmaceuticals, hormones, and other organic wastewater contaminants in U.S. streams, 1999–2000: a national reconnaissance. *Environmental Science and Technology* 36:1202–1211.

Kümmerer K. (2001) *Pharmaceuticals in the environment-source, fate, effects and risks*. Springer, Berlin.

Li H. S. (2000) *Experimental Techniques and Principles of Plant Physiology and Biochemistry*. Beijing. Higher Eduation Press. (in Chinese).

Liu F., Ying G. G., Tao R., Zhao L. L., Yang J. F., Zhao L. F. (2009) Effects of six selected antibiotics on plant growth and soil microbial and enzymatic activities. *Environmental Pollution* 157:1636-1642.

Sarmah A. K., Meyer M. T., Boxall A. B. A. (2006) A global perspective on the use, sales, exposure pathways, occurrence, fate and effects of veterinary antibiotics (VAs) in the environment. *Chemosphere* 65(5):725-759.

Schlüsener MP, Bester K (2006) Persistence of antibiotics such as macrolides, tiamulin and salinomycin in soil. Environmental Pollution 143:565-571.

Shi Y., Zhao M. Q., Wang J., Song Y. F. (2010) Effect of antibiotics from organic manure on protected vegetable soil-plant system: a review. *Journal of Agro-Environment Science* 29(supl.):240-244.

Thiele-Bruhn S. (2003) Pharmaceutical antibiotic compounds in soils: a review. *J Plant Nutr Soil Sci* 166:145–167.

In: Transgenic Plants
Editors: Shen Y. Zhu and Jiang L.Hu
ISBN 978-1-62257-245-8

Chapter 5

GROWTH RESPONSES OF BT AND NON-BT COTTONS TO SOIL PHOSPHORUS, COPPER AND CADMIUM LEVELS

***Lian Feng Du*[1] *and Wenke Liu*[2,*]**
[1]Institute of Plant Nutrition and Resources, Beijing Academy of Agriculture and Forestry Sciences, Beijing, China
[2]Institute of Environment and Sustainable Development in Agriculture, Chinese Academy of Agricultural Sciences, Key Lab for Agro-Environment and Climate Change, Ministry of Agriculture, Beijing, China

ABSTRACT

Two glasshouse experiments were conducted to investigate the growth differences of Bt- and non-Bt cotton on soils applied with various phosphorus (P), copper (Cu) and cadmium (Cd) levels. The results showed that high P level had no significant effect on shoot dry weight, root length and root dry weight of non-Bt cotton compared with low P treatment. However, high P level increased the shoot dry weight, root length, root dry weight and specific root length of Bt cotton. There was no difference between biological indices except specific root length between Bt cotton and non-Bt cotton under high P condition. Soil Cd and Cu addition inhibited the growth of both Bt and non-Bt cotton, with a

* Corresponding author; E-mail: liuwke@163.com.

significant decrease in shoot height, shoot dry weight and root dry weight. However, non-Bt cotton presented higher shoot height and shoot dry weight at the higher Cd and Cu levels.

Keywords: Bt cotton, cadmium, copper, phosphorus, soil adaptability

INTRODUCTION

Genetically modified crops (GMCs) are widely planted worldwide, and the global area of GMCs has continued to grow during the past fifteenth consecutive year since they were released in 1996. In 2010, there were 25 countries that had accepted and planted GMCs, with 148 million hm^2 total acreage (James 2010). Bt transgenic crops (BTCs) are major species of GMCs with insecticidal trait, and Bt transgenic cultivars of the first three species of GMCs (soybean, maize and cotton) have being developed. After inserting Bt genes, BTCs expressed Bt insecticidal proteins that could kill some halmful insects. However, Bt toxins also could be released into soil ecosystems, particularly in rhizosphere (Rui et al., 2005; Icoz and Stotzky 2008). Therefore, ecological risk assessment on soil ecosystems, especially rhizosphere microecosystem, had been regarded as important issue (Icoz and Stotzky 2008, Liu 2010).

Influences of Bt genes insertion on biological properties and physiological characteristics of BTCs are key issues should be clarified, which determine how to develop a strategy to make full use of insecticidal traits by optimizing agronomic measures, e.g. managment of irrigation and fertilization. Previous studies showed that Bt cottons demonstrated different biological properties and physiological characteristics compared with their parental lines, such as Bt insecticidal protein expression level (Dong and Li 2007), nitrogen metabolism (Chen et al. 2004), rhizosphere exudation of Bt insecticidal proteins (Rui et al., 2005; 2006), as well as regulation of environmental factors (Chen et al. 2005). However, little information about the growth responsive differences of BTCs and their parental cultivars to soil nutrient condition, e. g. soil phosporus (P) level, and to contaminated soils with heavy metals, e. g. copper (Cu) and cadmium (Cd). In this paper, two experiments were conducted to examine the growth responses of Bt and non-Bt cotton to soil P, Cu and Cd levels.

MATERIALS AND METHODS

Plants and Soils

Cotton cultivars used in the two experiments are Zhongmiansuo 41 (Bt cotton cultivar) and its parental line Zhongmiansuo 23 (non-Bt cotton cultivar). Bt cotton was transferred into bivalent genes, including Bt (*Bacillus thuringiensis*) and CpTI (cowpea trypsin inhibitor) genes. The former was purchased in market, and the latter was presented by Professor Jinjie Cui of Cotton Research Institute, Chinese Academy of Agricultural Sciences. Seeds were sterilized, and germinated under 30°C in dark incubator. Uniform germinated seeds were selected for experiments.

Soil of experiment 1 were collected from the vegetable field of Haidian District, Beijing. The soil used in experiment one contained 69.4mg/kg available nitrogen, 28.7mg/kg Olsen-P, 129.4mg/kg available potassium and pH 8.02 (water/ soil=2.5:1, w/w). Soil and building sand were air-dried and sieved through with 2 mm mesh, then mixed thoroughly for use. In order to lower available P concentration in soil, pure building sand was mixed with soil at ratio 1:2 (w/w). Basic fertilizer was applied, including nitrogen 200 mg (ammonium nitrate), potassium 80 mg (potassium sulfate), magnesium 5mg (magnesium sulfate heptahydrate), iron 5 mg (ferrous sulfate heptahydrate), copper 5mg (copper sulfate pentahydrate), zinc 5 mg (heptahydrate zinc sulphate) in per kilogram soil-river sand mixture. Soil used in experiment 2 was collected from vegetable field of Miyun County, Beijing. Organic matter content 10.5 g/kg, pH (water/soil=2.5:1) 7.4, total nitrogen 0.62 mg/kg, Olsen-P 51 mg/kg, available potassium 102.5 mg/kg, available nitrogen 51 mg/kg. Soil were air-dried and sieved through with 2 mm mesh, then mixed thoroughly for use. Basic fertilizer, nitrogen 100 mg (calcium nitrate), P 50 mg (potassium dihydrogen phosphate), potassium 100 mg (potassium sulfate), magnesium 5mg (magnesium sulfate heptahydrate), iron 5 mg (ferrous sulfate heptahydrate), copper 5mg (copper sulfate pentahydrate), zinc 5 mg (heptahydrate zinc sulphate) was also applied in per kilogram soil.

Experimental Design

In experiment 1, four treatments were desined, including two cotton cultivars (Bt-cotton and non-Bt cotton) and two soil P levels (20 mg and 60 mg P per kg soil). Each treatment replicated three times, 12 pots in total. Experiment began on March 15, 2006 in glasshouse of Institute of Environment and Sustainable Development in Agriculture, Chinese Academy of Agricultural Sciences. Sixty days after planting, cotton plants were harvested for determination. In experiment 2, ten treatments were designed, including two cotton cultivars and two Cu and Cd concentrations. A control without Cu and Cd addition was set up. Each treatment replicated three times. Two soil Cu concentrations are 100 and 150 mg/kg soil. Two Cd concentrations are 20 and 50 mg/kg soil. Cu was added in soil in form of copper sulfate pentahydrate (analytic grade), while Cd was added in form of sulfate cadmium octahydrate (analytic degrade). Each pot contained 1.8 kg soil and sand mixture. The experiment began on Nov. 11, 2009, and ended on Feb. 1, 2010. After emergency, two emergency, three seedlings were left.

Determination Methods

Shoot and root of cotton plants were harvested separately, and then rinsed cleanly with tap water and distilled water. Root length and specific root length were measured, then dried in oven under 70℃ for 48 hours, and weighed.

Statistics Analysis

Analysis of variance was conducted for both of the experiments for all the characteristics measured using Proc ANOVA in SAS. The differences between treatments were tested for significance using LSD.

RESULTS

Growth Responses of Bt Cotton and Non-Bt Cotton to SOIL P Levels

High P level had no significant effect on shoot dry weight, root length and root dry weight of non-Bt cotton compared with low P treatment (Table 1). However, high P treatments increased the shoot dry weight, root length, root dry weight and specific root length of Bt cotton. There was no difference between biological indices except specific root length between Bt cotton and non-Bt cotton under high P condition.

Growth Responses of Bt Cotton and Non-Bt Cotton to Soil Cd Levels

Compared with the controls, soil Cd addition inhibited the growth of both Bt and non-Bt cotton, with a significant decrease in shoot height, shoot dry weight and root dry weight (Table 2). However, non-Bt cotton presented higher shoot height and shoot dry weight at Cd50 treatments.

Table 1. Biological indices of two cotton varieties under two soil P levels

Treatments	Shoot dry weight (g/pot)	Root length(cm)	Root dry weight (g/pot)	Specific root length (m/g)
Non-BtP20	1.12a	7.6a	0.32a	30.3b
Non-BtP60	1.21a	10.9a	0.30a	43.3a
BtP20	0.58b	3.3b	0.13b	31.7b
BtP60	1.18a	8.0a	0.29a	35.5a

Note: In the same column, the different letters following the averages represent significant difference at $P<0.05$. Bt represents Bt cotton, while non-Bt stands for non-Bt cotton. P20 and P60 represent 20 mg/kg and 60mg/kg P treatments. The same below.

Table 2. Biological indices of two cotton varieties under two soil Cd levels

Treatments	Shoot height (cm)	Shoot dry weight (g/pot)	Root dry weight (g/pot)
Non-Bt0	10.4a	0.61a	0.15a
Non-BtCd50	8.4b	0.44ab	0.09c
Non-BtCd100	6.6bc	0.26c	0.06d
Bt0	10.7a	0.67a	0.13b
BtCd50	7.9bc	0.41b	0.09c
BtCd50	5.6c	0.27c	0.05de

Note: Cd50 and Cd100 represent 50 mg/kg and 10mg/kg Cd treatments.

Table 3. Biological indices of two cotton varieties under two soil Cu levels

Treatments	Shoot height (cm)	Shoot dry weight (g/pot)	Root dry weight (g/pot)
Non-Bt0	10.4a	0.61a	0.15a
Non-BtCu50	6.9b	0.28b	0.05cd
Non-BtCu100	6.6b	0.27b	0.06c
Bt0	10.7a	0.67a	0.13b
BtCu50	6.2b	0.11c	0.09cd
BtCu100	6.6b	0.12c	0.05d

Note: Cu50 and Cu100 represent 50 mg/kg and 10mg/kg Cu treatments.

Growth Responses of Bt Cotton and Non-Bt Cotton to Soil Cu Levels

Compared with the controls, soil Cu addition inhibited the growth of both Bt and non-Bt cotton, with a significant decrease in shoot height, shoot dry weight and root dry weight (Table 3). However, non-Bt cotton presented higher shoot dry weight at Cu50 and Cu100 treatments.

DISCUSSION

Although physiological characteristics of Bt cotton had been substantially revealed (Chen et al. 2004, Rui et al. 2005, 2006, Chen et al. 2005, Dong and Li 2007), the soil adaptablity of Bt cotton to soil fertility, particularly mineral nutrient levels and soil heavy metal contamination had not investigated. Present data showed that there were response differences in growth to soil P level. Previous reports and our data suggested that field management of Bt cotton should differ from their parental cultivars to improve yield or expression of Bt insecticial proteins. Our results showed that Bt cotton were more sensitive to low P condition compared with non-Bt cotton, which means that Bt cotton may need more available soil P. Previous literature showed that nitrogen metabolism of Bt cotton was higher than non-Bt cotton cultivars (Chen et al. 2004). Therefore, high P requirement of Bt cotton may be attributed to the increased P metabolism caused by the improved nitrogen metabolism strength for expression of Bt insecticidal proteins. Variation of soil P requirement of Bt cotton may influence some rhizosphere processes and microorganisms that related soil P level, e.g. arbuscular mycorrhizal fungi (Liu and Du 2008, Liu 2010).

Present data also showed that there were growth response differences in growth to soil Cu levels but Cd. Bt cotton and non-Bt cotton presented no growth differences on soil with various Cd levels, while Bt cotton tolerated higher soil Cu level than non-Bt cotton. High Cu tolerance of Bt cotton means that it can be cultivated on some Cu contaminated soils. Furthermore, Bt cotton may have potential for bioremediation for Cu contaminated soil. In conclusion, Bt gene insertion may altered P nutrition property and Cu tolerance of Bt cotton. Special management strategy should be made to optimize the management of irrigation and fertilization.

ACKNOWLEDGMENTS

I want to thank the financial support of the Open Fund of Key Laboratory of Crop Nutrition and Fertilization (2009), Ministry of Agriculture.

REFERENCES

Chen, D. H., Ye, G. Y., Yang, C. Q., Chen, Y. C. & Wu Y. K. (2004). Effect after introducing Bacillus thuringiensis gene on nitrogen metabolism in cotton. *Field Crops Research*, 87, 235–244.

Chen, D. H., Ye, G. Y., Yang, C. Q., Chen, Y. C. & Wu Y. K. (2005). The Effects of high temperature on the insecticidal properties of Bt Cotton. *Environmental and Experimental Botany*, 53, 333-342.

Dong, H. Z. & Li, W. J. (2007). Variability of endotoxin expression in Bt transgenic cotton. *Journal of Agronomy & Crop Science*, 193, 21-29.

Icoz, I. & Stotzky, G. (2008). Fate and effects of insect-resistant Bt crops in soil ecosystems. *Soil Biology & Biochemistry*, 40,559–586.

James, C. (2010). Global Status of Commercialized Biotech/GM Crops: 2010. ISAAA Brief No. 42. ISAAA: Ithaca, NY.

Liu, W. K. (2010). Do genetically modified plants impact arbuscular mycorrhizal fungi? *Ecotoxicology*, 19(2):229-238.

Liu, W. K. & Du, L. F. (2008). Interactions between Bt transgenic crops and arbuscular mycorrhizal fungi: a new urgent soil ecology issue in agro-ecosystems. *Acta Agriculturae Scandinavica Section B-Soil and Plant Science*, 58(2):187-192.

Rui, Y. K, Yi, G. X., Zhao, J, et a1. (2005). Changes of Bt toxin in the rhizosphere of transgenic Bt cotton and its influence on soil functional bacteria. *World Journal of Microbiology & Biotechnology*, 21:1279-1284.

Rui, Y. K., Yi, G. X., Guo, J., Guo, X., Luo, Y. B., Wang, B. M., Li, Z. H. (2006). Transgenic cotton could safely be grown since CpTI toxin rapidly degrades in the rhizosphere soil. *Acta Agriculturae Scandinavica Section B-Soil and Plant Science*, 57:122–125.

In: Transgenic Plants
Editors: Shen Y. Zhu and Jiang L.Hu

ISBN 978-1-62257-245-8

Chapter 6

THE APPLICATION OF SITE-SPECIFIC RECOMBINATION SYSTEMS FOR BIOSAFETY AND GENOME MANIPULATION IN THE PRODUCTION OF TRANSGENIC PLANTS

Yuan-Yeu Yau* and Ludmila Tyler
University of California-Berkeley, Albany, California, US

ABSTRACT

By enabling the development of new crop varieties with increased nutritional content, disease resistance, and stress tolerance, plant biotechnology plays an important role in addressing the food needs of the world's population. According to ISAAA Brief 39-2008, the area of land used globally for cultivating transgenic crops grew from 114.3 million hectares in 2007 to 125 million hectares in 2008. In 2007, the direct global farm income benefit from biotech crops was $10.1 billion (Brooke, 2009). So far, the majority of transgenic crops have been produced either by *Agrobacterium*-mediated transformation or by biolistic transformation. The gene controlling a trait of interest is usually delivered along with a selectable marker gene (such as an antibiotic-resistance gene) for later

* Corresponding author: Plant Gene Expression Center, USDA-ARS & Department of Plant & Microbial Biology, University of California-Berkeley, 800 Buchanan St., Albany, CA 94710.

selection of the transgenic plants from the bulk of non-transformed plants. After serving this purpose, the selectable marker gene is no longer needed. However, this approach for producing transgenic crops has raised concerns from both the regulatory and public sectors over the presence of selectable marker genes in the food supply or their possible escape into the environment through the outcrossing of transgenic crops to related species. Therefore, research in plant transgenesis continues to focus on ways to generate transgenic crops free of selectable marker genes. Several approaches for producing selectable-marker-free transgenic crops have been reported. Among these approaches, one of the most powerful is the application of microbial site-specific recombination systems to the transformation technology. This article will discuss the promise of transgenic breeding for future agriculture, the public concerns over genetically modified plants, and the utilization of site-specific recombination systems (including the novel ParA system recently described by our laboratory) for the production of marker-free transgenic plants. The use of site-specific recombination systems to tailor controlled transgene expression in plants and to genetically engineer plants for breeding programs will be discussed, as well.

Conventional Breeding and the Benefits of Transgenic Crop Production

The principles of genetics have been used for years to improve plant performance, especially yield and quality. In its early form, the application of genetics occurs through rough, non-systematic phenotypical or "visual" selections usually conducted by the farmers. Local farmers select and save the best-looking or highest-yielding plants and the largest seeds for future growing seasons. Plant lines resulting from this method are called "landraces" in the scientific community.

As genetics becomes better understood, advanced breeding methods with more complicated and systematic strategies are gradually developed and employed to increase yield in various crops. These methods include tailoring the selection schemes for specific crops. Conventional breeding is born. Plant breeders try to generate elite varieties in breeding programs by accumulating favorable alleles from within the gene pool or from cross-compatible wild relatives (Poehlman and Sleper, 1995). The practice usually involves sexually crossing two or more lines and selecting the lines with superior genotypes, which combine the desired traits and high heritability. The most prominent, successful example of conventional breeding is the Green Revolution, which started in the 1940s and continued into the 1950s and 1960s. The Green Revolution is often attributed to the American scientist and Nobel laureate (Peace Prize, 1970) Dr. Norman E. Borlaug. He originally worked to improve wheat by crossbreeding new disease-resistant (*e.g.* resistant to wheat rust), high-yielding, and dwarf wheat varieties in Mexico. The effort was a huge success. Mexico was, for the first time, able to produce more wheat than its own citizens needed and exported wheat by the 1960s.

The successful case of wheat in Mexico helped the breeding technologies spread worldwide, to be adopted by many countries, including the US and nations in Asia. The improvements in wheat were quickly replicated in other grains. For example, the development and release of a high-yielding, semi-dwarf variety of rice known as IR-8 from the International Rice Research Institute (IRRI) in the Philippines helped avert massive famine in India in the early 1960s. The IR-8 'magic rice' spread throughout Asia in the following decades, leading to the rice 'Green Revolution.' IR-8 was bred to include a "short stature" gene. The reduced height of IR-8 plants caused them to exhibit lodging-resistance, responsiveness to fertilizer inputs, and improved yields through increases in the harvest index (grain-straw ratio) (Khush, 1999). The short stature of IR-8 rice is now known to be due to a mutant gibberellin-synthesis gene, *sd1* (Sasaki *et al.*, 2002). This gene encodes an oxidase enzyme which is involved in the biosynthesis of the plant growth hormone gibberellin (GA). GA is one of the most important determinants of plant height. Because of the Green Revolution, global food production expanded dramatically, preventing hunger for millions of people (Borlaug, 2002). This type of traditional plant breeding is still commonly practiced today around the world.

Although it is an important tool in increasing plant performance and has improved many traits in different crops in the past, conventional breeding has some limitations. These limitations arise from the fact that genes for traits of interest can only be transmitted to the next generation by sexually crossing plants which bear these genes: *(1)* Crossing is difficult for plant species with primarily asexual modes of reproduction. Although some techniques can overcome this obstacle, they tend to be challenging and have limited applicability. Protoplast fusion, for example, requires a suitable method for protoplast production and is therefore not available for many plant species. *(2)* Sometimes even in species which reproduce sexually, a crossing barrier prevents crosses between different species from being successful. *(3)* When two plants are crossed, 50% of each parental genome will go into the next generation, which means that many traits – including undesirable ones – will exist in the offspring.

A prolonged process called backcrossing can be employed to recover the elite line's genomic makeup. However, a backcrossing program usually last several generations. This translates into years for *annual* crops (where the plant usually germinates, grows, flowers, and dies within a year), let alone for *biennial* or *perennial* crops. For biennial crops, a plant takes two years to complete its biological lifecycle: the plant grows vegetatively during the first year, then undergoes a *vernalization* process during a period of cold temperatures (winter), flowers in the spring of the following year, sets seeds, and finally dies. For perennial crops, a plant grows more than two years via a root-stock.

After recombinant DNA technology became available during the first half of the 1970s (Cohen *et al.*, 1973), genetic engineering emerged as a powerful new tool for modifying genomes, including those of plants. Genetic engineering and genetic transformation overcome the limitations of conventional breeding. Genetic engineering physically "removes" a gene by cloning it (PCR-amplifying it and cutting and pasting it into a DNA vector) from one organism – virtually any organism including bacteria – and transfers the gene into a plant by genetic transformation to create a transgenic plant with a new trait. Because sexual crossing is not needed, this method circumvents the "crossing barrier" which occurs between plants in conventional breeding programs. In addition to this advantage, the merits of transgenic breeding include, but are not limited to, the following: (1) Genes for new traits can be acquired from different species, including bacteria and animals, and (2) a novel gene can be created for new use(s) in a technique called "gene shuffling." Various gene fragments can be "shuffled" and engineered together to form a novel gene and then genetically transformed into plants for new function(s).

For example, Wulff *et al.* have generated a great variety of novel *Hcr9* (homologues of *C. fulvum* resistance gene *Cf-9*) shufflants *in vitro* by using members of the conserved *Hcr9* gene family for a gene shuffling experiment. Restriction-enzyme-cut fragments of DNA were used to generate a shufflant library. Some of these shufflants showed novel specificities and were found to act as auto-activators of the hypersensitive response, a plant defense mechanism, in *Nicotiana* species (Wulff *et al.*, 2004).

PRODUCTION OF TRANSGENIC PLANTS

1. Vector Construction

To start a genetic transformation experiment, recombinant DNA technology is used to build a construct based, for example, on a bacterial plasmid. For *Agrobacterium*-mediated transformation of plants, a gene of interest and a selectable marker gene are inserted between a pair of specialized DNA sequences, the left border (LB) and right border (RB) of the *Agrobacterium tumefaciens* transferred DNA (T-DNA). Constructs built without LB and RB sequences can be delivered into the plant nuclear genome by several methods described below, including the gene gun method and polyethylene glycol (PEG)-mediated protoplast transformation methods.

2. Genetic Transformation of Plants

There are several ways to genetically transform plants. They are: (1) *Agrobacterium*-mediated genetic transformation, (2) biolistic or particle bombardment, nicknamed "gene-gun" transformation, (3) electroporation of tissues or protoplasts, (4) microinjection, (5) polyethylene glycol (PEG)-mediated protoplast transformation, (6) virus-mediated transformation, and (7) whisker-mediated transformation. *Agrobacterium tumefaciens* is a soil-borne pathogen which infects many higher plant species and causes Crown Gall disease. *A. tumefaciens* contains a plasmid called the Ti, or tumor-inducing, plasmid. The bacterium has the ability to transfer and integrate a small part of the Ti plasmid (the T-DNA) into the genome of an infected plant. The proteins and mechanisms involved in the transfer and integration of T-DNAs have been discussed in great detail in many publications. An understanding of these processes has allowed scientists to engineer T-DNAs in order to design vectors for delivering genes of interest into higher plants. Particle bombardment uses a high-pressure gas (*e.g.* helium) to accelerate DNA-coated gold microparticles (0.6-1.6 μm in diameter) into plant tissues in a particle bombardment device such as the *Bio-Rad* PDS-1000/He™ system (*Bio-Rad* laboratories, Inc.). Electroporation uses pulses of high-voltage electricity to induce transient pores in cell membranes and thus allow for the entry of DNA molecules. Microinjection injects DNA molecules into a plant cell using a finely drawn micropipette needle. PEG-mediated protoplast transformation involves digesting plant cell walls with enzymes (cellulase and macerozyme) to produce protoplasts before PEG-treating the protoplasts for genetic transformation. Viral transformation proceeds by packaging the desired genetic material into a suitable plant virus and allowing this modified virus to infect the plants. Whisker-mediated genetic transformation involves the use of silicon carbide (SiC) whiskers (microfibers, 10-80 μm long and 0.6 μm in diameter) to deliver the DNA molecules into a plant's genome. The first report of plant tissue transformation utilizing this last method was published by Kaeppler *et al.* (1990). Transformed plant cells were generated by adding a DNA/fiber mixture to packed cells from maize or tobacco suspension cultures and vortexing for 60 seconds. Among these methods, *Agrobacterium*-mediated and gene-gun-mediated genetic transformation are the most popular for plants. *Agrobacterium*-mediated transformation had most often been used for dicots (plants with two cotyledons), while the gene-gun method was mostly used for monocots (rice, wheat, and other plants with a single cotyledon), although, nowadays, improved protocols allow several monocots (such as rice, wheat, maize, barley and sorghum) to be routinely transformed with *Agrobacterium* (see review by Cheng *et al.,* 2004).

3. Selection of Transgenic Plants

Because the transformation efficiency is less than optimal for many plant species, one of the most important steps in the production of transgenic plants is the ability to distinguish a few transgenic plants from the bulk of non-transgenic plants. To achieve this, researchers usually deliver a construct containing both the gene of interest and a selectable marker gene. Once this construct is integrated into the plant genome, the protein expressed from the selectable marker gene can, for example, detoxify a specific antibiotic or herbicide.

Using this mechanism, scientists can select the genetically transformed plant cells on a medium containing the antibiotic or herbicide. Non-transformed plants are susceptible to the selection agent and begin to die, while a resistant transgenic plant expressing the selectable marker gene will survive (Figure 1). Approximately fifty marker genes have reportedly been used for transgenic plant research or crop development (Miki and McHugh 2004). According to Miki and McHugh's investigation, the top three most commonly employed selectable marker genes are a kanamycin resistance gene, a hygromycin resistance gene, and a phosphinothricin resistance gene.

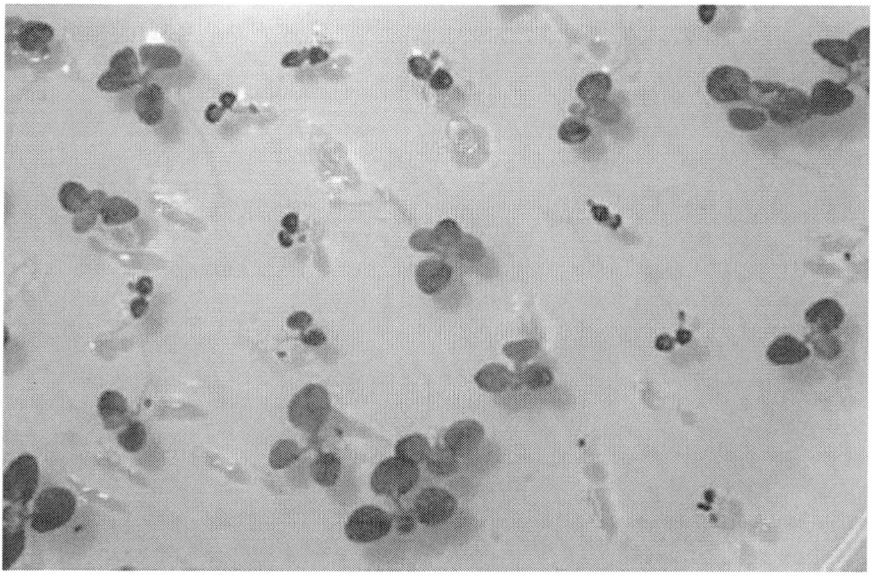

Figure 1. Tobacco seedlings on selection medium, MS + hygromycin (20 μg/mL). Resistant plantlets grow well in the medium, whereas susceptible plantlets remain small and fail to establish root systems.

The kanamycin resistance gene is derived from the bacterial amino-glycoside 3'-phosphotransferase II (APH [3'] II, E.C. 2.7.1.95) gene, also known as the neomycin phosphotransferase II (*npt*II) gene. The NPTII enzyme catalyses the ATP-dependent phosphorylation of the 3'-hydroxyl group of the amino-hexose portion of certain amino-glycosides including neomycin, kanamycin and geneticin. First, the *npt*II gene from *E. coli* is used to construct a binary vector for plant genetic transformation. Then, transformed plant tissues are selected by placing the tissues on a medium, such as MS (Murashige-Skoog) medium, containing kanamycin. Transgenic tissues expressing the *npt*II gene will survive kanamycin selection and grow in the medium. The concentration of kanamycin used depends on the plant species. For model plants such as *Arabidopsis* and tobacco, the concentration of kanamycin in the medium is usually 50 μg/mL for *Arabidopsis* transgenic seed selection and 100 μg/mL for tobacco transgenic seed selection or 200 μg/mL for tobacco transformed-leaf-tissue selection.

Hygromycin B is an aminocyclitol antibiotic inhibitor of protein synthesis with a broad spectrum of activity against prokaryotes and eukaryotes. The *E. coli* gene *aphIV* [also called *hph* or *hpt*, coding for hygromycin phosphotransferase (HPT, E.C. 2.7.1.119)] confers broad-spectrum resistance on prokaryotic and eukaryotic cells, including cells from higher plants. In the absence of the *aphIV* gene, hygromycin B is very toxic to plants. Sterile, liquid hygromycin can be purchased from a commercial supplier and added directly to the medium for transgenic plant selection. Members of the genus *Streptomyces* produce hundreds of antibiotics, one of which is *bialaphos* (also known as PPT, for phosphinothricin, or glufosinate). Bialaphos inhibits glutamine synthetase (GS), the key enzyme in the nitrogen assimilation pathway. The toxicological characteristics of glufosinate have made it a popular herbicide, and it has been commercialized under the names Basta® and Liberty® (http://www.patentlens.net/daisy/Phosph/ext/navaggregator/navaggregator). Antibiotic-producing *Streptomyces* strains have evolved mechanisms to avoid the toxicity of their own products. Such strains often contain modifying enzymes which can inactivate the antibiotics, such as an enzyme that detoxifies glufosinate. Thompson *et al.* (1987) have isolated the gene encoding this enzyme from *Streptomyces hygroscopicus.* The gene – named *bar* (for bialaphos resistance) – encodes a phosphinothricin acetyl transferase (PAT). Numerous researchers have used the *bar* gene as a selectable marker, cloning it into DNA constructs for plant genetic transformation. Transgenic plants expressing PAT can survive and grow in medium supplemented with the bialaphos compound. Scientists can also select *bar*-gene-expressing transgenic plants by spreading the bialaphos solution directly onto the seedlings or leaves of the plants.

CONCERNS ABOUT THE USE OF SELECTABLE MARKER GENES

Many questions have been raised regarding the safety of transgenic crops, both for the consumer and the environment. The fact that selectable marker genes remain in the genomes of genetically modified plants has raised both regulatory and public concerns, which include: (1) If plant products consumed by humans as food contain antibiotic resistance genes, will these products increase the incidence of antibiotic-resistant bacterial strains? (2) If seeds or pollen grains contain herbicide-resistance genes, will spreading of the transgenic seeds and crossing to wild relatives via the transgenic pollen ("*gene flow*") contaminate the environment by creating new, aggressive weeds which cannot be eliminated by herbicide application? Gene flow of an herbicide-resistance transgene has been studied in Canada in herbicide-resistant canola fields (Warwick *et al.*, 2003). In this study, researchers observed that gene flow occurred from commercial, herbicide-resistant *B. napus* to natural, wild populations of *B. rapa*, if the plants were grown close to each other. However, the researchers also found that barrier crops can be planted as "absorbers" to prevent gene flow.

Due to the concerns described above, before the release of genetically modified plants from laboratories for field trials and commercialization, intensive studies and tests regarding potential impacts on human health and the environment need to be done, and regulatory processes need to be set. Although studies with mice revealed that *npt*II degraded rapidly in a simulated gastric environment and was unlikely to cause an allergic response (Fuchs *et al.*, 1993), and although gene flow can be carefully handled and monitored to prevent it, much uncertainty still exists. Besides, studies of the impact of other marker genes, such as an herbicide-resistance gene, on human or animal health are rare, and the outcomes remain unclear.

These risks may, however, be alleviated by removing the selectable marker genes once they are no longer needed. Therefore, the production of selectable-marker-gene-free transgenic crops is becoming an extremely attractive alternative. The holy grail of current transgenic plant research has been to develop strategies that can remove the vector backbone and selectable marker genes, after their use, from the final transgenic product. In addition, removing the selectable marker gene would allow researchers to re-use the same selectable marker, without the need to utilize multiple markers. Consequently, several approaches have been developed to generate selectable-marker-gene-free transgenic crops.

Strategies for Producing Marker-Gene-Free Transgenic Plants

As reviewed by Hohn *et al.* (2001) and Darbani *et al.* (2007), there are several approaches to producing marker-free transgenic plants. These approaches are:

1) Avoid using selectable marker genes in the transformation process for producing genetically modified (GM) crops. Theoretically, because the transformed plants carry a transgene, it should be possible to distinguish them from un-transformed plants using molecular tools

such as PCR. However, a project like this would be extremely labor-intensive, time-consuming, and costly. Without selection, the majority of the screened plants would be un-transformed. This approach is impractical even in today's laboratories with the availability of automated analysis and with a PCR protocol which directly uses plant tissues for PCR amplification, eliminating the need for isolating genomic DNA. (Kits for performing PCR from plant material include the Phire® Plant Direct PCR kit sold by Biolabs, www.finnzymes.com.).

2) Use a "co-transformation and segregation"-based strategy to remove selectable marker genes. One example of this approach is to build the selectable marker gene and the transgene of interest (such as a trait gene) into two separate constructs which are co-transformed into plant tissues by either *Agrobacterium*-mediated or biolistic genetic transformation. Resistant transgenic plants surviving selection are screened for the presence of the gene of interest. By selecting for the transgene of interest and not the marker gene in the segregating population of the subsequent generation, one can obtain transgenic lines with the transgene of interest but without the marker gene. Under this strategy for producing marker-free plants, co-transformation using *Agrobacterium* – rather than particle bombardment – stands out as the more-often-used method; this is likely due to the simplicity and efficiency of *Agrobacterium*-mediated transformation. Many variations of this approach, referred to as "2 T-DNA" transformation, have been described (Huang *et al.*, 2004). The co-bombardment method is analogous to the "2 T-DNA" *Agrobacterium*-mediated transformation method. However, there are only a handful of case studies utilizing the co-bombardment method to produce marker-free transgenic plants. This may be because bombardment tends to deliver multiple copies of transgenes into a single locus to form a complex insertion. In general, studies on producing marker-free plants through bombardment are limited, although unlinked integrations of transgenes have been reported, and a low percentage of marker-free plants were recovered after segregating away the unlinked marker gene. By using "cassette DNA (not the whole plasmid)" and "limiting the amount of input DNA," Prakash *et al.* (2009) devised a novel method for producing marker-free transgenic corn through co-bombardment. The so-called "cassettes" were DNA fragments isolated from plasmids via appropriate restriction-enzyme

digestions. The bombardments were performed with 2.5 ng of the *npt*II-gene cassette per shot and 15 ng of the gene-of-interest cassette per shot, amounts which are much lower than the 0.6-1 μg/shot suggested by the manufacturer. With this approach, the researchers identified 28 marker-free corn plants from the progeny of 103 R_0 plants containing both co-transformed constructs. There are disadvantages to the "co-transformation and segregation"-based approach: (1) It is time-consuming and labor-intensive. (2) This method can only be applied to sexually propagated plants whose genomes undergo recombination and segregation, not to vegetatively propagated plants. (3) In addition, if the two constructs become closely linked to each other when they are introduced into plants, then the chances of finding two separated transgenes (the marker gene and the transgene of interest) in the segregating population will be dramatically reduced.

3) Employ a "transposon"-based strategy for marker-gene removal. The maize transposable element *Ac* has the ability to move to new locations within a genome (Fedoroff, 1989). According to a previous study, about 10% of the *Ac* elements that are excised do not reinsert and therefore disappear from the genome or reinsert into a sister chromatid that is subsequently lost during somatic segregation. Ebinuma *et al.* (1997) employed the *Ac* element to remove the isopentenyl transferase gene (*ipt*), which was used for visually distinguishing transgenic plants from non-transgenic plants by identifying the *ipt* "shooty" phenotype. In the experiment, a T-DNA was constructed to contain an *Ac* element linked to a 35S promoter-*ipt* gene-terminator marker-gene cassette, and a gene-of-interest cassette. The idea is that when the *Ac* element transposes, the *ipt* gene transposes along with the *Ac* element and may eventually be lost with the *Ac* element, resulting in a T-DNA which only contains the gene-of-interest cassette. By six months post infection, the researchers had obtained marker-free transgenic tobacco plants – from which *Ac* had disappeared – from 4.8% of the transgenic clones.
4) Remove selectable marker genes via intrachromosomal homologous recombination of two bacteriophage λ *attP* sites. Usually, using intrachromosomal homologous recombination (ICR) between two homologous sequences to induce transgene deletion in plants is not efficient, due to the low frequency of ICR. According to previous studies, ICR between closely linked repeated sequences occurs at a

rate of approximately 10^{-6} (Puchta and Hohn, 1996). However, Peter Meyer's group has devised an ICR strategy which is based on the recombination of the *attP* region of bacteriophage λ but which does not require the presence of recombinase and other helper proteins (Zubko *et al.*, 2000). In this study, a construct was built to contain two 250-bp λ phage attachment sites (or *attP*) as homologous repeats flanking a selectable marker gene. The construct was transformed into tobacco plants, and higher rates of ICR-induced selectable-marker-gene deletion were detected in different parts of two of eleven lines. There were no detectable deletion events in the other nine lines. This result was attributed to position effect, in this case, the influence which the integration region of the individual transgene has on the efficiency of recombination. In parallel, the researchers observed that ICR is not always associated with precise homologous recombination between the two *attP* regions; larger deletions (in which regions outside the *attP* sequences were lost, as well) were also detected among the deletion events in the two lines. Illegitimate recombination was proposed as the underlying mechanism for the majority of the deletion events. It is unclear why the rates of ICR-induced DNA-fragment (selectable-marker-gene) excision were higher than expected. Puchta (2000) has suggested several possibilities, including that the increased rates may be caused by T-DNAs integrating into genomic double-stranded breaks (DSBs), and the only known way to increase homologous–and simultaneously, illegitimate– recombinations by several orders of magnitude is to induce DSBs. However, except for this paper, apparently no other laboratories have reported using a similar approach to remove selectable markers. It would be interesting to explore if other *attP*-like sites from other site-specific recombination systems can be used for this same purpose.

5) Utilize zinc finger nucleases (ZFNs) for marker gene removal. The ZFNs are fairly new molecular tools for manipulating plant genomes, including mutating and adding genes. Genomic integrity is very important for maintaining a healthy cell, and cells must repair damaged DNA with high fidelity. One of the most dangerous DNA lesions is a DSB, which can lead to cell death or to mutation. When eukaryotic chromosomes suffer DSBs, the cellular DNA repair systems are activated to repair these breaks through both homology-dependent (homologous recombination or HR) and non-homology-dependent (nonhomologous end-joining or NHEJ) mechanisms.

Repair of DSBs by NHEJ generates localized mutations by causing small deletions and insertions at the break site. The ZFNs, originally termed hybrid restriction enzymes, are fusions of zinc-finger-based DNA recognition modules to endonuclease domains (Kim *et al.*, 1996). These enzymes are engineered to contain a zinc finger DNA-binding domain (composed of 3 zinc fingers) and the nonsequence-specific DNA-cleavage domain from the *Fok*I type II restriction endonuclease (Fn domain) (Porteus and Carroll, 2005). The cleavage domain of ZFNs must dimerize in order to cut DNA; therefore, efficient cleavage requires two zinc-finger-binding sites in close proximity. Because the zinc finger domain can be engineered to recognize a wide range of novel DNA sequences, a specific ZFN can be designed, produced, introduced and directed to a specific genomic sequence for cleavage. ZFNs have been widely used by genomic engineers to introduce DSBs at endogenous loci in eukaryotic systems to generate gene mutations (due to a small deletion and insertion of sequences at the breaks by the NHEJ repair mechanism) and to achieve gene replacement (through the homologous recombination repair mechanism). For instance, Petolino *et al.* (2010) demonstrated that a *GUS* reporter gene flanked by two ZFN cleavage sites was deleted from a stably transformed plant by crossing it with a second plant expressing a corresponding ZFN gene. In this study, ZFN binding sites flanking the targeted gene were pre-integrated into the plant genome; however, since ZFNs can be designed to bind and cleave a wide range of endogenous DNA sequences, these results constitute a first test of a general strategy for targeted gene deletions (Petolino *et al.*, 2010).

6) Exploit site-specific recombination to remove selectable marker genes. Site-specific recombination systems, which are discussed extensively below, have gained in popularity over the years, as a method for marker gene removal. These genetic systems contain three basic components— the recombinase protein and two recombination sites, which are either identical or dissimilar, for example *attP* or *attB*. These components are essential for site-specific integration or excision between the two molecules which contain the recombination sites. Researchers have cloned the DNA sequences for these three components and used them to build constructs for various applications in transgenic plant production. Dale and Ow (1991) conducted pioneering work to demonstrate that the Cre-*lox* site-specific

recombination system was capable of excising unnecessary selectable marker genes from transgenic plants. In the two decades since then, numerous papers have been published using Cre-*lox* or other newly-identified site-specific recombination systems for marker gene removal in plants. The common feature of these studies is the production of transgenic plants with a selectable marker gene flanked by two recombination sites oriented in the same direction. Upon expression of the recombinase, a site-specific recombination event loops out the marker gene placed between the recombination sites.

The Sources of Site-Specific Recombination Systems

Site-specific recombination systems are commonly found in prokaryotes and lower eukaryotes such as the unicellular yeasts. The original biological functions of site-specific recombination systems are to integrate or excise bacteriophage into or out of the host genome, to maintain plasmid copy number during replication, to "switch" gene expression, to alter cell-surface components in some bacteria, to generate the final products of transposon transposition, and to accelerate the replication process by inverting the convergent replication forks in autonomous plasmids (Stark *et al.*, 1992).

Based on evolutionary and mechanistic relatedness, the recombinase super-family is composed of two fundamental groups, the *tyrosine* recombinases and the *serine* recombinases, named after the active amino acid residues that forms covalent protein-DNA linkages in the reaction intermediates (Grindley *et al.*, 2006). Virtually all of the site-specific recombination systems fall into these two groups. Serine recombinases are subdivided into the small recombinases, which catalyze only intramolecular strand exchange (excision and/or inversion), and large recombinases, which catalyze both inter-and intramolecular strand exchange (integration and excision) (Smith and Thorpe, 2002).

Site-specific recombination is characterized by the following three features: (1) recombination occurs at specific DNA sequences (known as "recognition sites") on the interacting DNA molecules, (2) recombination is conservative, with the reaction resulting in no gain or loss of nucleotides, and (3) strand exchange occurs in small regions (with homologous DNA sequences) within the recognition sites.

The simplest site-specific recombination systems comprise two identical recognition sites and a single-polypeptide recombinase protein that catalyzes the recombination event (Sadowski, 1993). These systems include Cre-*lox* from the *E. coli* bacteriophage P1 (Sauer and Henderson, 1988), FLP-*FRT* from the 2-μ cycle yeast *Saccharomyces cerevisiae* (Golic and Lindquist, 1989), and R-*RS* from *Zygosaccharomyces rouxii* (Onouchi *et al.*, 1991); Cre, FLP and R are the recombinases, and *lox*, *FRT,* and *RS* are the respective DNA recognition sites. More complex site-specific recombination systems may have recombination sites that are dissimilar in sequence and may require different recombinases and accessory protein(s) (Grindley *et al.*, 2006).

Modes of Action of Site-Specific Recombination

Depending on the initial arrangement of recognition sites on the involved molecule(s), site-specific recombination can lead to four possible outcomes: translocation, integration, excision, and inversion of DNA molecules (Figure 2 and Figure 3). Recombination between recognition sites *in trans* causes a reciprocal translocation of two linear DNA molecules, while co-integration occurs if at least one of the two molecules is circular (Ow, 2002).

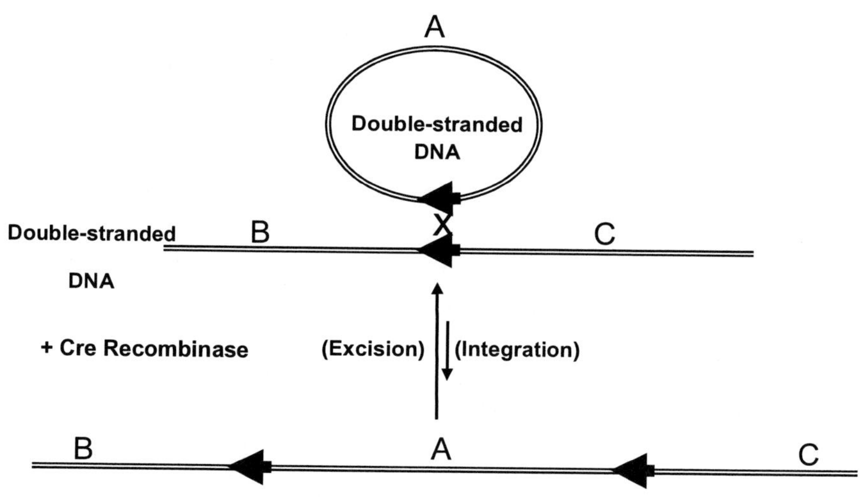

Figure 2. Cre-protein-mediated integration and excision reactions between two DNA molecules carrying *loxP* sites (short, dark arrows). The reactions are freely reversible, but excision is kinetically favored over integration.

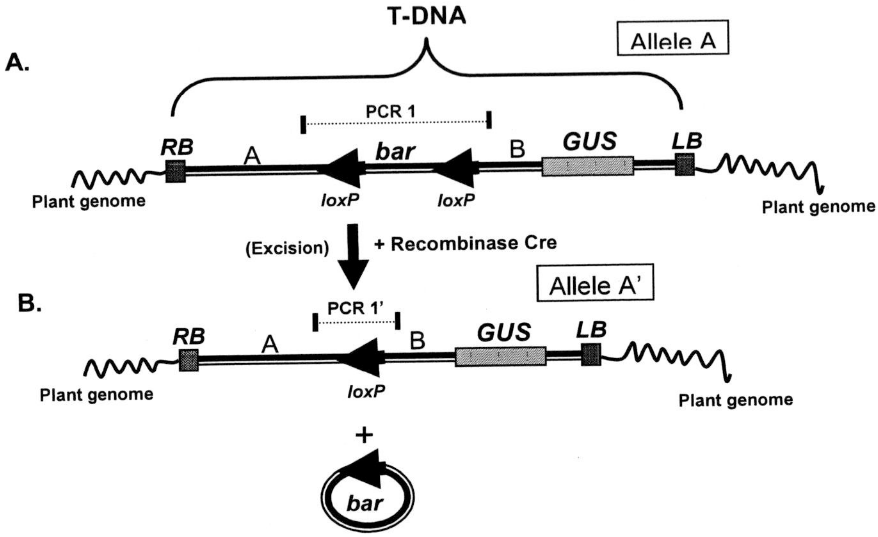

Figure 3. Cre-mediated removal of a selectable marker gene. A T-DNA fragment containing a selectable marker gene, *bar*, and a reporter gene, *GUS*, is integrated into the plant genome (allele A) (A). The *bar* gene is flanked by a pair of *loxP* sites (in the same orientation, arrowheads). The *bar* gene is excised from the T-DNA after a site-specific recombination event catalyzed by the Cre recombinase, yielding the new allele A' (B). The plants with allele A' can be identified and distinguished from plants with allele A by using PCR. (Contrast the sizes of the products for PCR1 *vs*. PCR1'.). RB: right border of the T-DNA, LB: left border of the T-DNA, A: DNA fragment "A," B: DNA fragment "B," *bar*: *bar* selectable marker gene, *GUS*: β-glucuronidase reporter gene.

Excision of intervening DNA results from recombination between two recognition sites oriented in the same direction *in cis*. Inversion events result from recombination between two recognition sites oriented in opposite directions (Grindley *et al*., 2006). The simplest site-specific recombination systems Cre-*lox*, FLP-*FRT* and R-*RS*, belong to the small tyrosine group. Within these systems, the genetic cross-over between the two recognition sites regenerates a single site which is identical in nucleic acid sequence to the parental sites and releases the intervening DNA fragment. Because of the identical nature of the two recognition sites, the recombination reactions (co-integration and excision) are fully reversible, although the intramolecular reaction (excision) is kinetically favored over the intermolecular reaction (co-integration) (Figure 2). A second group of site-specific recombination systems contains β-*six* (Diaz *et al*., 2001), the bacterial γδ (Tn1000) transposon-encoded resolvase/invertase sub-family site-specific DNA recombination system γδ-*res* (Schwikardi and Dröge, 2000), CinH-*RS2* (Kholodii G, 2001; Thomson and Ow, 2006), and ParA-*MRS* (Gerlitz *et al.*, 1990; Thomson *et al.*, 2009); β, γδ, CinH, and ParA are small serine recombinases, and *six, res*, *RS2*, and *MRS* are the respective DNA recognition sites. The natural function of the γδ-*res* system is to resolve, through recombination, the co-integrate structure that is generated during transposition (Stark *et al.*, 1992). Recombination mediated by these small serine recombinases also regenerates a single product site of the same sequence as the parental sites. However, as mentioned earlier, the small serine recombinases catalyze intra-molecular excision; due to conformational stain, they cannot facilitate intermolecular integration.

A third group of site-specific recombination systems is represented by phiC31 (Thomason *et al.*, 2001), TP901-1 (Stoll *et al.*, 2002), R4 (Olivares *et al.*, 2001), and Bxb1 (Kim *et al.*, 2003; Keravala *et al.*, 2006; Thomson and Ow, 2006). These large serine recombinases (to distinguish from the small serine site-specific recombinases) act on two recognition sites, *attB* and *attP* (*b*acterial and *p*hage *att*achment sites), that differ in sequence to yield hybrid sites known as *attL* and *attR*.

These recombinases can facilitate an excision, inversion or integration reaction but cannot reverse the reaction, because the recognition-site sequences of *attB* and *attP* change to the sequences of *attL* and *attR*. The hybrid sites *attL* and *attR* are no longer substrates for the recombinase proteins. Reversing the reaction is only possible through the addition of a second protein, an excisionase (Ghosh *et al.*, 2006; Thorpe *et al.*, 2000). The phiC31 integrase is a 68-kDa, single-polypeptide, bacteriophage serine recombinase. It catalyzes site-specific recombination to allow integration of the *Streptomyces* phage phiC31 into the host chromosome (Smith and Thorpe, 2002).

Cre, FLP and R are also called *bi-directional* recombinases because the recombination reactions (co-integration and excision) are fully and freely reversible. These particular recombinases are capable of excision, inversion, and integration. phiC31 is an example of a *uni-directional* site-specific recombination system. Once co-integration occurs between two molecules through *attB* and *attP* site-specific recombination, the integrated molecule cannot be freely excised. To date, only members of the tyrosine recombinase family have shown bi-directional activity, while both tyrosine and serine recombinase families have shown uni-directional functionality. Furthermore, within the serine recombinase family, there are members which specialize in deletion reactions and cannot catalyze inversion or integration reactions.

A Brief History of the Study of the Cre-lox Site-Specific Recombination System and its Application in Eukaryotic DNA Excision

Cre-*lox* is the best-known and most widely investigated site-specific recombination system, as well as the foremost system for site-specific recombination studies in eukaryotic cells. Dr. Brian Sauer is credited with discovering this system (Sauer, 1987), but many workers from diverse fields have found creative ways to utilize Cre-*lox* recombination in manipulating mammalian and plant genomes. Cre stands for "causes recombination," and *loxP* stands for "locus of crossing over in P1." The native *loxP* site is composed of an 8-bp spacer region flanked by two identical 13-bp inverted repeats (arm regions). P1 is maintained in the *E. coli* genome as a single-copy plasmid. The role of the Cre recombinase is to resolve dimers of P1 that arise after replication, in order to allow partitioning of the two P1 molecules at cell division.

In 1981, Sternberg and his colleague found that "a site (called *loxP*) must be present in both partners in the recombination in order for recombination to occur, and a P1 gene (called *cre*), whose product is necessary for recombination, must also be present" (Sternberg and Hamilton, 1981). In 1982, Hoess *et al.* reported the detailed nucleotide sequence of the *loxP* site, began to clarify which DNA sequences are necessary for *loxP* recombination, and implicated a region of dyad symmetry as being important (Hoess *et al.*, 1982). In 1983, Abremski *et al.* (1983) demonstrated that when Cre recombinase was used *in vitro*, the products of recombination between *loxP* sites on a supercoiled DNA substrate were present as free supercoiled circles, *i.e.* topologically unlinked products. In 1984, Abremski and his colleague described the purification and characterization of the Cre recombinase protein. In an *in vitro* study, they observed that intramolecular recombination (two *loxP* sites on the same DNA molecule) is considerably more efficient than intermolecular recombination (Abremski and Hoess, 1984). In 1985, Hoess and his co-worker reported the mechanism of strand cleavage and exchange in the Cre-*lox* site-specific recombination system, locating the point at which strand exchange takes place with one radioactively labeled and one unlabeled *loxP* site (Hoess and Abremski, 1985). In 1987, Sauer was the first to demonstrate that the prokaryotic Cre-*lox* site-specific recombination system of phage P1 could function in an efficient manner in a lower eukaryote, the yeast *Saccharomyces cerevisiae*. This study demonstrated that a prokaryotic recombinase (Cre) can enter a eukaryotic nucleus and efficiently and accurately catalyze excision and recombination at specific DNA sites (*loxP* sites) to remove the fragment between them (Sauer, 1987). This finding opened the door for scientists to start using prokaryotic site-specific recombination systems for eukaryotic genome manipulation. In 1988, Sauer and Henderson showed that Cre can be functionally expressed in cultured mouse cells to catalyze excisive recombination of extrachromosomal DNA (Sauer and Henderson, 1988). In 1989, Sauer and Henderson showed that the Cre protein can promote DNA recombination at *loxP* sites stably resident on a mammalian chromosome, resulting in the release of a 2.5 kb yeast DNA fragment flanked by two directly repeated *loxP* sites (Sauer and Henderson, 1989). In 1990, David Ow's group was the first to demonstrate the functioning of the Cre-*lox* system in plants (Dale and Ow, 1990). The group then reported, in 1991, that the Cre protein can precisely and conservatively (without loss or alteration of nucleotides in the recombination sites) catalyze excision events in the tobacco genome to remove a selectable marker gene (Dale and Ow, 1991). Since then, numerous publications have employed the Cre-*lox* site-specific recombination system for mammalian and plant genome manipulation, including excising selectable marker genes from transgenic plants. In parallel, other site-specific recombination systems were discovered and tested in eukaryotic cells.

APPLICATIONS OF SITE-SPECIFIC RECOMBINATION SYSTEMS FOR PLANT GENOME MANIPULATION

1. Site-Specific Recombination Systems Are Used to Delete Selectable Marker Genes in Plants

The most straightforward application of site-specific recombination in plant cells is the removal of undesired DNA segments, especially unwanted selectable marker genes from transgenic plants. After the discovery that the Cre-*lox* system functioned well in plants (Dale and Ow, 1990), site-specific recombination systems have been used in a number of different ways to remove selectable marker genes, for example in transgenic tobacco (Dale and Ow 1991; Bayley *et al.*, 1992; Gleave *et al.*, 1999) and rice (Hoa *et al.*, 2002). This approach for removing unneeded selectable marker genes has the benefit, as described earlier in this chapter, of reducing concerns raised by regulatory groups and consumers about the introduction of new genes. Site-specific excision has also allowed recycling of the selectable marker genes for multiple rounds of genetic modification in plants (Dale and Ow, 1991). For example, the kanamycin resistance gene (*npt*II) is by far the most popular selectable marker used in transgenic crop production. However, it can be difficult to use the same selectable marker to introduce a second gene of interest into a transgenic plant that already contains an *npt*II resistance gene in the genome. Although other selectable markers exist, only a limited number are of practical use. Therefore, it is desirable to develop a system for removing the selectable marker gene after there is no need for its presence, thus permitting re-use of the marker and alleviating safety concerns.

1.1. Constitutively Expressing a Recombinase Gene for Selectable-Marker-Gene Excision

In initial studies using site-specific recombination systems to remove unwanted DNA marker genes in plants, two main strategies were employed: line crossing (target line x recombinase-expressing line) and re-transformation (two rounds of transformation). (1) For line crossing, target lines are developed by transforming plants with a T-DNA construct containing a gene of interest and a selectable marker gene which is flanked by recognition sites. The target lines are then screened via Southern analysis for single-copy transgenes and for precise construct structure (no rearrangements, truncations, or similar events occuring after transformation). In the meantime, recombinase expressing lines are developed by genetically transforming plants and screening for lines which express the recombinase at high levels. The characterized target lines are then crossed to the characterized recombinase-expressing lines. The F_1 seeds derived from the cross are harvested, germinated and screened for the presence of both the target gene and the recombinase gene. Progeny of the F_1 plants are then tested for germinal transmission of the final product of excision. (2) For the re-transformation strategy, after the target lines are generated and screened, they are subjected to another round of genetic transformation, this time with a recombinase-expressing gene construct (for which a different selectable marker is needed). The re-transformed lines are screened for the presence of the recombinase-expressing gene. Transgenic lines with the gene are allowed to set seeds, and plants derived from these seeds are tested for germinal transmission of the marker-gene-excised, final transgenic product. The un-needed recombinase gene can then be removed from the marker-free lines by genetic segregation. Although marker-free transgenic crop plants can be produced in laboratories using these two processes, there are some disadvantages: the crossing strategy involves generating and then crossing target lines and recombinase-expressing lines, selecting for complete excision which is heritable, and finally segregating away the recombinase gene. This strategy is restricted to sexually propagated species, and some crops, such as potato, are only propagated asexually. For other crops, long generation times make crossing schemes impractical. On the other hand, the re-transformation method is lengthy. There are also reports that overexpression of the recombinase gene may cause abnormalities in the transgenic plants (Coppoolse *et al.*, 2003). Similarly, undesired effects following the overexpression of site-specific recombinases in mammalian cells have been documented (Buchholz, 2009). Loonstra *et al.* (2001) reported that Cre expression caused numerous chromosomal aberrations and markedly reduced proliferation in cultured mammalian cells. Also, Naiche *et al.* (2007) observed that Cre activity causes widespread apoptosis and lethal anemia during embryonic development, and Liu *et al.* (2009) found that phiC31 integrase induces a DNA damage response and chromosomal rearrangements in human adult fibroblasts.

1.2. Transiently Expressing a Recombinase Gene for Selectable-Marker-Gene Excision

There are ways to avoid the possibility of causing aberrant developmental phenotypes due to a constitutively overexpressed, stably integrated recombinase gene, as well as to bypass the need for crossing the recombinase gene in and out. These approaches include: (1) The recombinase can be transiently expressed from a DNA construct which, instead of inserting into the plant genome, is eventually degraded and removed from the plant cells. (2) The recombinase gene can be placed under an inducible promoter and arranged with the marker gene in such a way that, when the recombinase gene is activated, it can auto-excise its own coding sequence, as well as that of the marker gene. This section (1.2) discusses transient expression, while the next section (1.3) examines inducible expression.

Agrobacterium T-DNA vectors and plant virus vectors can serve as vehicles for delivering recombinase genes into plant cells for transient, high-level expression of recombinase proteins, without integration of the recombinase genes into the genome. There have been reports of successful, site-specific-recombinase-mediated selectable-marker-gene excisions through transient expression of Cre, with both *Agrobacterium* T-DNA vectors (Kopertekh and Schiemann, 2005) and plant-RNA-virus-based vectors

(Kopertekh *et al.*, 2004; Jia *et al.*, 2006) being used as vehicles to deliver the *cre* gene.

Virus-based vectors are particularly suitable for temporary recombinase expression in plants, since the foreign genes delivered by viruses into the infected plant and amplified by viral replicase do not integrate into the plant genome (Koprowski and Yusibov, 2001). Jia and co-workers described the use of a movement-function-improved m30B viral vector [including a modifying movement protein (MP) gene from a well-developed plant virus vector, the Tobacco Mosaic Virus (TMV) 30B vector], to transiently express Cre recombinase and thus remove a selectable marker gene from transgenic tobacco. *Agrobacterium* carrying the TMV vector was pressure-infiltrated, with a needle-less syringe, into plant tissues containing an excision construct consisting of *loxP* sites flanking the *npt*II selectable marker. The authors obtained selectable-marker-free transgenic plants with 34% excision efficiency. The progeny inherited the marker-free genotype from the regenerated parental plants and were also virus-free, since TMV is not transmitted to seeds. A further examination by Southern analysis confirmed that no stably integrated *cre* gene was detected in the genomes of the marker-free transgenic plants. Other plant-virus-based vectors have also reportedly been used to delete marker genes via transient *cre* gene expression. This includes using a Potato Virus X (PVX)-based vector to transiently express Cre to remove a selectable marker gene from transgenic *Nicotiana benthamiana* (Kopertekh *et al.*, 2004).

Transient recombinase expression was also demonstrated in plants by biolistic. In an attempt to produce Cre-mediated site-specific integration in corn, Srivastava and co-worker (2001) co-bombarded a *cre*-gene construct and an integration vector into the target line callus. The data showed that 60% of the single-copy-transgene-inserted plants lacked the recombinase gene, indicating that *cre* gene expression was transient. Additionally, polyethylene glycol (PEG)-mediated transformation can deliver a recombinase gene on a plasmid into protoplasts for transient expression of the recombinase protein (Albert and Ow, 2000; Yau *et al.*, 2010). Yau *et al.* used this method to integrate a gene of interest into a pre-determined locus in the tobacco genome. For this study, the authors co-transformed two constructs, the "integration construct" containing a Bxb1 *attB* site and the *bxb1* (or Bxb1-recombinase-expressing) construct, into protoplasts derived from a tobacco target line. The target line had previously been genetically transformed to embed a Bxb1 *attP* site into the genome. The *bxb1* construct provides recombinase activity for the *attB*-containing integration plasmid to recombine with the chromosomal *attP* site. By selecting only for the selectable marker carried on the integration construct, several putative integration lines were obtained. The integration lines were examined further to determine whether they contained the Bxb1 recombinase gene. No *bxb1* gene was detected, indicating that none had stably integrated into the genome and that the Bxb1 recombinase protein which facilitated integration of the integration construct into the embedded *attP* site in the target line was transiently expressed.

1.3. Induced Expression of a Recombinase Gene for Selectable-Marker-Gene Excision

In order to further control the excision process to remove the selectable marker at a certain time, recombinase genes driven by various inducible promoters have been developed. These include *heat-shock-inducible* promoters and *chemical-inducible* promoters. By activating the promoters with inducers (heat or chemicals), the expression of the recombinase gene can be tightly controlled. To shorten and simplify the process of excising selectable markers, various groups have designed a one-step auto-excision strategy for DNA removal using inducible promoters. In this strategy, the marker gene, the recombinase gene, and the gene(s) of the interest are cloned into a single construct. The recombinase gene under the control of an inducible promoter and the marker gene are placed into a cassette flanked by two recognition sites oriented in the same direction, while the gene of interest is placed outside the region flanked by the recognition sites. After transformation and molecular analysis, the transgenic plants are treated with inducers to initiate expression of the recombinase (*e.g.* Cre) to delete the region between the recognition sites (*e.g. loxP*); the recombinase and the selectable marker gene are both removed. Recent studies using this method have been reviewed by Gidoni *et al.* (2008). For example, a heat-shock-regulated, Cre-*lox*-mediated, site-specific excision of extraneous DNA in transgenic tobacco plants was reported (Liu *et al.*, 2005). In this study, the *Arabidopsis* heat-shock-responsive promoter *HSP81-1* (Takahashi *et al.,* 1992) was used to control the expression of Cre recombinase. In nearly all cases, the selectable markers (*bar* and *npt*II), the Cre gene, and other unneeded components located between two 34-bp *loxP* sites were efficiently removed after heat-shock treatments: Testing a total of 115 leaves (from 10 lines) revealed that the excision frequency was 95.7%. This strategy of using a heat-shock-inducible promoter to control timely expression of the recombinase gene has also been applied to *Arabidopsis* (Hoff *et al.*, 2001), maize (Zhang *et al.*, 2003), tobacco (Wang *et al.*, 2005), potato

(Cuellar *et al.*, 2006) and aspen (Fladung *et al.*, 2010) for site-specific recombination events.

Chemically regulated, site-specific DNA excision to remove the selectable marker gene in transgenic plants via the site-specific recombination systems was also reported in *Arabidopsis*) [by Cre-*lox*] (Zuo *et al.*, 2001), aspen [by FLP-*FRT*] (Matsunaga *et al.*, 2002), rice [by Cre-*lox*] (Sreekala *et al.*, 2005), tomato [by Cre-*lox*] (Zhang *et al.*, 2006; Zhang *et al.*, 2009) and tobacco [by FLP-*FRT*] (Woo *et al.*, 2009). In the study by Zuo *et al.*, an estrogen-receptor-based transactivator termed XVE [for the bacterial repressor LexA (X), the acidic transactivating domain of VP16 (V), and the regulatory region of the human estrogen receptor (E) (Zuo *et al.*, 2000)] and the Cre-*lox* system were used in combination. In transgenic plants, β-estradiol-activated XVE stimulated expression of the Cre protein under the control of the target promoter; this resulted in Cre-mediated site-specific recombination events and deletion of the unnecessary DNA The XVE system was subsequently adopted by several groups for producing marker-free transgenic rice (Sreekala *et al.*, 2005) and tomato (Zhang *et al.*, 2006; Zhang *et al.*, 2009). Woo *et al.* (2009) reported auto-excision of selectable marker genes from transgenic tobacco via a stress-inducible FLP/*FRT* site-specific recombination system. In this system, an oxidative-stress-inducible promoter [a peroxidase (POD) promoter] was fused to the FLP recombinase gene. Hydrogen peroxide (H_2O_2) was used to induce the recombinase gene, and the successful excision of selectable marker genes via hydrogen peroxide (H_2O_2)-regulated site-specific recombination was observed. Because heat-shock treatments and chemicals are required for recombinase activation for marker gene deletion, these treatments may be limited to certain plant species and may present complications for the transformation process (Li *et al.*, 2007). Other problems associated with these inducible systems include low promoter induction, which can lead to incomplete transgene excision, and background activity, which is due to unwanted leaky expression and leads to unintended excision. Leakiness in promoter activity among transgenic lines is common and has been reported by several laboratories who used chemical- or heat-shock-inducible promoters for recombinase gene expression (Gidoni *et al.*, 2008).

To prevent potential leakiness of the promoter, a 'double-lock' method was employed by Joubès *et al.* (2004) to tightly control a heat-shock-inducible promoter (*HSP18.2*) and thus avoid premature activation of the inducible system. This strategy combines the Cre-*lox* site-specific recombination system and a mammalian glucocorticoid receptor (GR) [specifically, the hormone-binding domain of the rat GR] in a single construct for selectable-marker-gene removal. The *cre* recombinase gene is fused to the GR gene, and this entire coding region is driven by a heat-shock-inducible promoter. Hence, the ligand-inducible recombinase is under the control of two inducers: Heat shock induction of the promoter provides regulation at the transcriptional level, and glucocorticoid activation of the Cre-GR fusion provides regulation at the post-transcriptional level. In the absence of glucocorticoids, the Cre-GR protein remains in the cytosol, but when chemical inducer is added, the fusion protein moves into the nucleus. To accomplish selectable marker deletion, stably transformed tobacco calli were first treated at 37°C for 2 hours to activate the *cre* gene. Subsequently, the calli were treated with a glucocorticoid compound, dexamethasone, for 24 hours to trigger translocation of the GR-fused Cre protein into the nucleus, where Cre excised sequences flanked by *loxP*. The researchers found that, in 93% of the calli, the DNA fragment (*egfp*) between the two *loxP* sites was excised through Cre-*lox*-mediated recombination. No deletion was detected when only chemical induction or only heat treatment was used, confirming the effectiveness of the "double-lock" approach.

1.4. Genetically Programmed Recombinase Gene Expression for Selectable Marker Excision

Although inducible promoters offer control over the timing of recombinase expression, the need for external induction and problems with efficacy make inducible promoters less than ideal. One issue concerns how to efficiently apply inducing chemicals to transgenic plants on the farm. Application methods used in the laboratory – where plant tissue or callus can be conveniently placed on medium supplemented with the inducer(s) in a sterile Petri dish – are not feasible in the field. Besides, the chemicals used for induction could negatively impact other plants on the farm. Pre-mature recombination events, due to recombinases expressed from leaky promoters, are another concern. Chimerism, and possibly cost, has also hampered widespread field applications of such advanced approaches. Therefore, it is more desirable to have an auto-excision event catalyzed by recombinase proteins which are driven by a genetically controlled, endogenous trigger.

Several labs have made progress in this direction: They have successfully utilized pollen-specific or tissue-specific promoters to drive recombinase gene expression for the genetically controlled excision of selectable markers (Mlynárová *et al.*, 2006; Li *et al.*, 2007; Luo *et al.*, 2007; Verweire *et al.*, 2007; Frédéric *et al.*, 2009). When a pollen-specific promoter is used to regulate recombinase gene expression, selectable marker gene removal becomes an integral part of the biology of pollen maturation. Mlynárová *et al.*

(2006) have used the tobacco NTM19 promoter to drive the Cre recombinase gene (*cre*), in order to auto-excise a *loxP*-flanked cassette embedded in transformed plants. The activity of the NTM19 promoter, which is highly specific to the microspore, is first detected in early uni-nucleate microspores and is absent from all vegetative tissues. The *loxP*-flanked cassette, in this case, was composed of the "NTM19 promoter-*cre* gene" DNA fragment and a selectable marker gene, connected to a gene of interest outside the *loxP* sites. Once the NTM19 promoter was activated during early pollen development, Cre -mediated site-specific recombination excised the DNA fragment located between the *loxP* sites. The system was highly efficient, with a failure rate of only 2 out of the 16,800 seeds which were derived from the selectable-marker-free tobacco pollen. Additionally, expression of the *cre* gene was tightly regulated; leakiness of the promoter was not observed (Mlynárová *et al.*, 2006). Verweire *et al.* (2007) also successfully demonstrated the use of *Arabidopsis* germline-specific promoters from the *APETALA1* and *SOLO DANCERS* genes to achieve genetically programmed, Cre-recombinase-mediated auto-excision. Marker-free progeny plants were efficiently obtained in this study, as well. More recently, Frédéric *et al.* (2009) evaluated seven different germline-related promoters for their suitability in regulating Cre expression in transgenic plants. Five out of the seven promoters were able to drive efficient Cre-mediated gene excision. The data also showed that use of these promoters resulted in much less variation in recombination frequency than previously reported for the 35S promoter.

2. Site-Specific Recombination Systems Are Used for Resolution-Based Single-Copy Transgenic Line Production

Agrobacterium-mediated transformation and particle bombardment are the two most widely used methods for plant transformation. Unfortunately, these methods of gene delivery lead to the random integration of transgenic material throughout the plant genome, and the number of transgenes that insert into the host chromosome cannot be determined in advance. Complex, multi-copy patterns in the genome add another concern, as they are often linked to gene silencing (Meyer and Saedler, 1996). Furthermore, evidence has indicated that DNA integration is often fragmented or concatenated at a single locus. When transgenic plants are produced through *Agrobacterium*-mediated transformation, the integration of multiple T-DNA copies into a single locus as direct or inverted repeats is fairly common (De Buck *et al.*, 1999). Moreover, biolistic (gene-gun) transformation tends to result in more complex integrations than *Agrobacterium*-mediated transformation does (Tobias *et al.*, 2007). It is possible to obtain a concatenated DNA array consisting of 3 to 500 or more transgene "monomers" at a single integration site (Dorer, 1997). These concatenated transgenes, especially when arranged as inverted repeats, are unstable due to the interaction of the repeated, homologous sequences within the genome; these circumstances can lead to variegated transgene expression, as seen in transgenic mice (Dobie *et al.*, 1997); transgene silencing, such as in transgenic plants (Wang and Waterhouse, 2000; De Buck *et al.*, 2001); and genetic instability from deleted sequences (Collick *et al.*, 1996). The expression level of the transgene does not necessarily correspond to the number of copies in the integrated array and must be determined experimentally by assaying mRNA and/or protein levels.

The formation of concatenated transgenes is a common problem in genomic manipulation. Therefore, it is desirable to develop a method for resolving these concatemers. Although the site-specific Cre-*lox* recombination system is mainly used to remove selectable marker genes, exposure of the modified genome to Cre recombinase has also been utilized to reduce the tandem arrays of transgenes at a single locus to a single integrated copy. This technique was originally shown to function in wheat by Srivastava and colleagues (1999). For this procedure, the transgenes or T-DNAs need to be engineered to contain one or two *loxP* sites. When a transgenic plant with a complex array of T-DNAs at a locus (regardless of the orientation and number) is crossed to a Cre-expressing plant, site-specific recombination between the outermost *loxP* sequences in the same orientation should resolve the multiple copies into a single T-DNA. Further exploiting the features of the Cre-*lox* system, Srivastava and Ow (2001) conducted an experiment to generate single-copy primary transformants. The authors hypothesized that co-bombarding a transgene flanked by oppositely-oriented *loxP* sites into maize calli along with a Cre-expressing construct would result in low copy numbers of the transgene insertion in the majority of the primary transformants, because Cre would resolve the concatemerized transgene copies before or after they were inserted into the plant genome. Indeed, 85% of the transformants had 1-2 transgene insertions. In another example, De Buck *et al.* (2007) designed a specialized T-DNA construct that contains two inversely-oriented *loxP* sites located inside and immediately adjacent to the T-DNA right and left borders, in order to produce a single-copy transgene insertion at a locus. This transgene reduction technique ensures more control over the system by resolving complex patterns of integration and allowing direct comparison between sites of integration and gene expression.

3. Site-Specific Recombination Systems Are Used in Breeding Programs for Hybrid Seed Production

Hybrids are the offspring derived from intercrossing between two inbred lines or among more than two inbred lines, *e.g.* four in the case of "double-crosses." In many crop species, hybrids are desirable because of superior agronomic characteristics, especially in terms of yield. Hybrids of most crops yield 10-30% more than their pure inbred parental lines (Mayo, 1980). This phenomenon is called "*heterosis* or hybrid vigor." Heterosis is the opposite of *inbreeding depression*, which occurs with increasing homozygosity. Some hypotheses, including the dominance and overdominance hypotheses, have been offered to explain the genetic basis of heterosis. The dominance hypothesis attributes the superiority of hybrids to the suppression of undesirable or deleterious recessive alleles from one parent by dominant alleles from the other parent. However, the heterosis effect disappears after the first (F_1) generation, so it is pointless for farmers to save seeds produced by a hybrid crop.

Since hybrid vigor is so powerful, breeders have continuously tried to develop new methods for producing hybrid seeds or crops to increase yield. This includes developing better inbred varieties, conducting intensive field trials of inbred varieties, and devising novel crossing schemes for producing hybrids. Corn and rice hybrids have been the most successful. However, because many important food crops are self-pollinating, producing hybrid seed often requires methods for promoting outcrossing. For example, for hybrid corn production, two inbred varieties can be planted in a field, in alternating rows. The tassels (male part) of one variety are removed to ensure that the second variety is the only pollen source in the field. Unfortunately, de-tasseling is labor-intensive and costly.

Another valuable method is to utilize cytoplasmic male-sterile (CMS) lines which produce no pollen and can serve as the female parent for hybrid crossing. For instance, breeders have sought and exploited CMS lines to produce hybrid rice. A male-fertility-restoring gene is usually crossed into the male parent used in hybrid seed production, so that the hybrid plants will be able to self-pollinate *(http://vivi.cornell.edu/* impact/individual/vivo/ individual16535). However, there are several drawbacks to CMS systems, such as, very few CMS lines are available, and it is hard to obtain good maintainer and restorer lines. Therefore, seeking alternative methods for producing hybrid crops has become one of the most important goals for crop breeders. Luo *et al.* (2000) have incorporated FLP/*FRT* site-specific recombination in designing a sterility-fertility restoration system for hybrid seed production. In this study, male-sterile lines were produced by transforming an antisense *Arabidopsis bcp1* gene construct into *Arabidopsis*. The *bcp1* gene is essential for pollen fertility in *Arabidopsis*. At the same time, the researchers also generated FLP-recombinase-expressing lines. All the progeny of crosses between the sterile lines and the FLP-recombinase-expressing lines produced normal siliques and seeds. These results demonstrated the utility of site-specific recombination for developing hybrids.

Gils *et al.* (2008) have also employed a phiC31-based system to develop a two-component hybrid seed production system to effectively generate and maintain the male-sterile parent, produce hybrid seeds, and fully restore fertility in the hybrid plants. Utilizing excision mediated by the phiC31 recombinase, the researchers exploited the functional interaction of two transgenic loci that are in the same position on two homologous chromosomes. Two transgenic fragments (A1 and A2) in the two loci were generated from a single DNA construct containing the A1-A2 components. The DNA construct (A1-A2), built with the phiC31 recombination sites *attP* and *attB*, was transformed into *Arabidopsis*. The transformed lines were then crossed to plants expressing phiC31 recombinase. After site-specific excision, A1-only or A2-only progeny were generated. By re-crossing these progeny, the researchers obtained the male-sterile lines and the maintainer lines.

4. Novel Site-Specific Recombination Systems Can Potentially Be Exploited for Plant Genome Manipulations

Although there are currently several microbial site-specific recombination systems available for research uses, some of the systems are patented by the private sector. It is therefore worthwhile to identify other efficient site-specific recombination systems for genome modification *in planta*. The new-found site-specific recombination systems can also be used together with already-existing systems to expand the toolkit for plant genome manipulations. For example, a plant can be engineered to contain both Cre and phiC31 site-specific recombination components in the genome for site-specific excision and integration purposes (Srivastava and Ow, 2004).

Recently, our group, under the leadership of David W. Ow, reported that four members of the small serine resolvase subfamily, CinH, ParA, Tn1721, and Tn5053, can catalyze site-specific excision in a lower eukaryote, yeast. Compared to Cre-*lox*, the estimated deletion rates for CinH, ParA, Tn1721, and Tn5053 were 91.8%, 93.3%, 54.8% and 72.6%, respectively (Thomson and Ow, 2006). ParA is a small serine resolvase from a plasmid operon, *parCBA*, which is responsible for the maintenance of the broad host range plasmids RK2 and RP4; ParA catalyzes the recombination of two identical 133-bp recognition sites known as *MRS*. The ParA site-specific recombination system has been used in genetic transformations to remove selectable marker genes from *Arabidopsis* (Thomson *et al.*, 2009). The ParA recombinase excised the *MRS*-flanked DNA, and the excision event was detected in subsequent generations in the absence of the ParA enzyme, indicating germinal transmission of the excision event. We also tested the phiC31 system for its ability and efficacy in removing a selectable marker gene via site-specific recombination in *Arabidopsis*. In our experiments, the phiC31 system has proven highly effective in removing – through site-specific excision – a transgenic, target sequence embedded in the *Arabidopsis* genome and in generating stably excised progeny plants that carry only the recombined target DNA of interest in the absence of the recombinase gene (Thomson *et al.*, 2010a). Likewise, Kempe *et al.* (2010) reported that the phiC31 site-specific recombinase could excise a transgene from wheat chromosomes, and that the recombinant loci were inherited by the subsequent generation. The other site-specific recombination systems, including Bxb1, CinH and β, are also being investigated for applications in the genomic engineering of crop plants (Gronlund *et al.* 2007; Thomson *et al.*, 2010b). Gronlund *et al.* (2007) reported using the β/*six* system to delete a spacer region that is flanked by directly oriented *six* recognition sites. However, the β/*six* system requires an accessory protein, a host factor, for recombination.

Final Remarks and Future Perspectives

Genetic transformation techniques have been used to engineer disease resistance and other traits into plants, generating GM crops. The acceptance of biotech crops has increased steadily, and they have becoming a trend in farming. The "ISAAA Brief 39-2008: Executive Summary" from the International Service for the Acquisition of Agri-Biotech Applications

(ISAAA) reported the following statistics for 2008: (1) the number of countries electing to grow biotech crops grew to 25, up from 6 in 1996, (2) the global hectarage of biotech crops increased to 125 million, and (3) the number of farmers growing GM crops reached 13.3 million globally. Amazingly, in 2008, 85% of the United States' maize crop, measured in hectares, was GM. Biotech crops have contributed to food security and the alleviation of poverty and hunger, as well as reducing impacts on the environment. In China, the cultivation of *Bt* cotton by small farmers has reduced insecticide use by 60% (ISAAA Brief 39-2008). Biotechnology cannot, and is not intended to, replace conventional breeding. Instead, biotechnology is complementary to conventional breeding; together, the two approaches promote progress in agriculture.

However, marker genes left behind in the process of generating transgenic plants have caused concerns among regulatory agencies and consumers. Producing marker-gene-free crops will alleviate these concerns. As this review outlines, many approaches have been employed to delete marker genes in transgenic plants, with good results. In particular, over the past 20 years, site-specific recombination systems have become one of the most popular tools for marker gene deletion. These systems have proven effective for manipulating plant genomes but will require fine-tuning to transition from the laboratory environment to farmers' fields.

REFERENCES

Abremski, K., Hoess, R. (1984) Bacteriophage P1 site-specific recombination: purification and properties of the Cre recombinase protein. *J. Biol. Chem.* 259: 1509-1514.

Abremski, K., Hoess, R., Sternberg, N. (1983) Studies on the properties of P1 site-specific recombination: evidence for topologically unlinked products following recombination. *Cell* 32: 1301-1311.

Albert, H., Ow, D.W. (2000) Cre-*lox* directed integration of transgenes into the tobacco genome. *Plant Molecular Biology Manual* N3: 1-13.

Abremski, K., Hoess, R., Sternberg, N. (1983) Studies on the properties of P1 site-specific recombination: evidence for topologically unlinked products following recombination. *Cell* 32: 1301-1311.

Bayley, C.C., Morgan, M., Dale, E.C., Ow, D.W. (1992) Exchange of gene activity in transgenic plants catalyzed by the Cre-*lox* site-specific recombination system. *Plant Mol. Biol.* 18: 353-361.

Borlaug, N.E. (2002) Feeding a world of 10 billion people: the miracle ahead. *In Vitro Cell, Dev. Biol.-Plant* 38: 221-228.

Brookes, G., Barfoot, P. (2009) Global impact of biotech crops: Income and production effects, 1996-2007. *Ag. Bio. Forum* 12: 184-208.

Buchholz, F. (2009) Engineering DNA processing enzymes for the postgenomic era. *Current Opinion in Biotechnology* 20: 383-389.

Cheng, M., Lowe, B.A., Spencer, T.M., Ye, X., Armstrong, C.L. (2004) Factors influencing *Agrobacterium*-mediated transformation of monocotyledonous species. *In Vitro Cell, Dev. Biol. Plant* 40: 31-45.

Cohen, S.N., Chang, A.C.Y., Boyer, H.W., Helling, R.B. (1973) Construction of biologically functional bacterial plasmids *in vitro*. *Proc. Natl. Acad. Sci. USA* 70: 3240-3244.

Collick, A., Drew, J., Penberth, J., Bois, P., Luckett, J., Scaerou, F., Jeffreys, A., Reik, W. (1996) Instability of long inverted repeats within mouse transgenes. *EMBO J.* 15: 1163-1171.

Coppoolse, E.R., de Vroomen, M.J., Roelofs, D., Smit, J., van Gennip, F., Hersmus, B.J.M., Nijkamp, H.J.J., van Haaren, M.J.J. (2003) Cre recombinase expression can result in phenotypic aberrations in plants. *Plant Mol. Biol.* 51: 263-279.

Cuellar, W., Gaudin, A., Solȯrzano, D., Casas, A., Nopo, L., Chudalayandi, P., Medrano, G., Kreuze, J., Ghislain, M. (2006) Self-excision of the antibiotic resistance gene *nptII* using a heat inducible Cre-loxP system from transgenic potato. *Plant Mol. Biol.* 62: 71-82.

Dale, E., Ow, D.W. (1990) Intra- and intermolecular site-specific recombination in plant cells mediated by bacteriophage P1 recombinase. *Gene* 91: 79-85.

Dale, E., Ow, D.W. (1991) Gene transfer with subsequent removal of the selection gene from the host genome. *Proc. Natl. Acad. Sci. USA* 88: 10558-10562.

Darbani, B., Eimanifar, A., Stewart, Jr., C.N., Camargo, W.N. (2007) Methods to produce marker-free transgenic plants. *Biotechnol. J.* 2: 83-90.

De Buck, S., Jacobs, A., Van Montagu, M., Depicker, A. (1999) The DNA sequences of T-DNA junctions suggest that complex T-DNA loci are formed by a recombination process resembling T-DNA integration. *Plant J.* 20: 295-304.

De Buck, S., Van Montagu, M., Depicker, A. (2001) Transgene silencing of invertedly repeated transgenes is released upon deletion of one of the transgene involved. *Plant Mol. Biol.* 46: 433-445.

Diaz, V., Servert, P., Prieto, I., Gonzalez, M.A., Martinez, A.C., Alonso, J.C., Bernad, A. (2001) New insights into host factor requirements for prokaryotic beta-recombinase-mediated reactions in mammalian cells. *J. Biol. Chem.* 276: 16257-16264.

Dobie, K., Mehtali, M., McClenaghan, M., Lathe, R. (1997). Variegated gene expression in mice. *Trends Genet.* 13: 127-130.

Dorer, D. (1997) Do transgene arrays form heterochromatin in vertebrates? *Transgenic Res.* 6: 3-10.

Ebinuma, H., Sugita, K., Matsunaga, E., Yamakado, M. (1997) Selection of marker-free transgenic plants using the isopentyl transferase gene. *Proc. Natl. Acad. Sci. USA* 94: 2117-2121.

Fedoroff, N.V. (1989) in *Mobile DNA*, eds. Douglas, E.B. and Martha, M.H. (Am. Soc. Microbiol., Washington, DC), pp. 375-411.

Fladung, M., Becker, D. (2010) Elimination of marker genes and targeted integration via FLP/*FRT* recombination system from yeast in hybrid aspen (*Populus tremula* L. x *P. tremuloides* Michx.). *Tree Genet. Genomics* 6: 205-217.

Frédéric, V.E., Verweire, D., Claeys, M., Depicker, A., Angenon, G. (2009) Evaluation of seven promoters to achieve germline directed Cre-*lox* recombination in *Arabidopsis thaliana*. *Plant Cell Rep.* 28: 1509-1520.

Fuchs, R.L., Ream, J.E., Hammond, B.G., Naylor, M.W., Leimgruber, R.M., Berberich, S.A. (1993) Safety assessment of the neomycin phosphotransferase II (NPTII) protein. *Biotechnology* 11: 1543-1547.

Gerlitz, M., Hrabak, O., Schwab, H. (1990) Partitioning of broad-host-range plasmid RP4 is a complex system involving site-specific recombination. *J. Bacteriol.* 172: 6194-6203.

Ghosh, P., Wasil, L.R., Hatfull, G.F. (2006) Control of phage Bxb1 excision by a novel recombination directionality Factor. *PLoS Biol.* 4: e186. doi:10.1371/journal.pbio.0040186

Gils, M., Marillonnet, S., Werner, S., Grutzner, R., Giritch, A., Engler, C., Schachschneider, R., Klimyuk, V., Gleba, Y. (2008) A novel hybrid seed system for plants. *Plant Biotechnol.* 6: 226-235.

Gidoni, D., Srivastava, V., Carmi, N. (2008) Site-specific excisional recombination strategies for elimination of undesirable transgenes from crop plants. *In Vitro Cell Dev. Biol.- Plant* 44: 457-467.

Gleave, A.P., Mitra, D.S., Mudge, S.R., Morris, B.A.M. (1999) Selectable marker-free transgenic plants without sexual crossing: transient expression of *cre* recombinase and use of a conditional lethal dominant gene. *Plant Mol. Biol.* 40: 223–235.

Golic, K.G., Lindquist, S. (1989) The FLP recombinase of yeast catalyzes site-specific recombination in the *Drosophila* genome. *Cell* 59: 499-509.

Grindley, N.D., Whiteson K.L., Rice P.A. (2006) Mechanisms of site specific recombination. *Annu. Rev. Biochem.* 75: 567-605.

Gronlund, J.T., Stemmer, C., Lichota, J., Merkle, T., Grasser, K.D. (2007) Functionality of the β/*six* site-specific recombination system in tobacco and *Arabidopsis*: a novel tool for genetic engineering of plant genomes. *Plant Mol. Biol.* 63: 545-556.

Hoa, T.T., Bong, B.B., Huq, E., Hodges, T.K. (2002) Cre/*lox* site-specific recombination controls the excision of a transgene from the rice genome. *Theor. Appl. Genet.* 104: 518-525.

Hoess, R.H., Abremski, K. (1985) Mechanism of strand cleavage and exchange in the Cre-*lox* site specific recombination system. *J. Mol. Biol.* 181: 351-362.

Hoess, R.H., Ziese, M., Sternberg, N. (1982) P1 site-specific recombination: Nucleotide sequence of the recombining sites. *Proc. Natl. Acad. Sci. USA* 79: 3398-3402.

Hoff, T., Schnorr, K.M., Mundy, J.A. (2001) Recombinase-mediated transcriptional induction system in transgenic plants. *Plant Mol. Biol.* 45: 41-49.

Hohn, B., Levy, A.A., Puchta, H. (2001) Elimination of selection markers from transgenic plants. *Curr. Opin. in Biotechnol.* 12: 139-143.

Huang, S., Gilbertson, L.A., Adams, T.H., Malloy, K.P., Reisenbigler, E.K., Birr, D.H., Snyder, M.W., Zhang, Q., Luethy, M.H. (2004) Generation of marker-free transgenic maize by regular two-border *Agrobacterium* transformation vectors. *Transgenic Res.* 13: 451-461.

Jia, H., Pang, Y., Chen, X., Fang, R. Removal of the selectable marker gene from transgenic tobacco plants by expression of Cre recombinase from a tobacco mosaic virus vector through Agroinfection. *Transgenic Res.* 15: 375-384.

Joubès, J., De Schutter, K., Verkest, A., Inzé, D., De Veylder, L. (2004) Conditional, recombinase-mediated expression of genes in plant cell cultures. *Plant J.* 37: 889-896.

Kaeppler, H. F., Gu, W., Somers, D. A., Rines, H.W., Cockburn, A.F. (1990) Silicon carbide fiber-mediated DNA delivery into plant cells. *Plant Cell Rep.* 9: 415-418.

Kempe, K., Rubtsova, M., Berger, C., Kumlehn, J., Schollmeier, C., Gils, M. (2010) Transgene excision from wheat chromosomes by phage phiC31 integrase. *Plant Mol. Biol.* 72: 673–687.

Keravala, A., Groth, A.C., Jarrahian, S., Thyagarajan, B., Hoyt, J.J., Kirby, P.J., Calos, M.P. (2006) A diversity of serine phage integrases mediate site-specific recombination in mammalian cells. *Mol. Gen. Gent.* 276: 135-146.

Kholodii G. (2001) The shuffling function of resolvases. *Gene* 269: 121-130.

Khush, G.S. (1999) Green revolution: preparing for the 21st century. *Genome* 42: 646-655.

Kim, A.I., Ghosh, P., Aarron, M.A., Bibb, L.A., Jain, S., Hatfull, G.F. (2003) Mycobacteriophage Bxb1 integrates into the *Mycobacterium smegmatis groELI* gene. *Mol. Microbiol.* 50: 463-473.

Kim, Y.G., Cha, J., Chandrasegaran, S. (1996) Hybrid restriction enzymes: zinc finger fusions to *Fok* I cleavage domain. *Proc. Natl. Acad. Sci. USA* 93: 1156-1160.

Kopertekh, L., Jűttner, G., Schiemann, J. (2004) PVX-Cre-mediated marker gene elimination from transgenic plants. *Plant Mol. Biol.* 55: 491-500.

Kopertekh, L., Schiemann, J. (2005) Agroinfiltration as a tool for transient expression of *Cre* recombinase *in vivo*. *Transgenic Res.* 14: 793-798.

Koprowski, H., Yusibov, V. (2001) The green revolution: plants as heterologous expression vectors. *Vaccine* 19: 2735-2741.

Li, Z., Xing, A., Moon, B.P., Burgoyne, S.A., Guida, A.D., Liang, H., et al. (2007) A Cre/loxP-mediated self-activating gene excision system to produce marker gene free transgenic soybean plants. *Plant Mol. Biol.* 65: 329-341.

Liu, H.-K., Yang, C., Wei, Z.-M. (2005) Heat shock-regulated site-specific excision of extraneous DNA in transgenic plants. *Plant Sci* 168: 997-1003.

Liu, J., Skjorringe, T., Gjetting, T., Jensen, T.G. (2009) PhiC31 integrase induces a DNA damage response and chromosomal rearrangements in human adult fibroblasts. *BMC Biotechnol.* 9: 31.

Loonstra, A., Vooijs, M., Beverloo, H.B., Allak, B.A., van Drunen, E., Kanaar, R., Berns, A., Jonkers, J. (2001) Growth inhibition and DNA damage induced by Cre recombinase in mammalian cells. *Natl. Acad. Sci. USA* 98: 9209-9214.

Luo, H., Lyznik, A., Gidoni, D., Hodges, T.K. (2000) FLP-mediated recombination for use in hybrid plant production. *Plant J.* 23: 423-430.

Luo, K., Duan, H., Zhao, D., Zheng, X., Deng, W., Chen, Y., Stewart Jr, C.N., McAvoy, R., Jiang, X., Wu, Y., He, A., Pei, Y., Li, Y. (2007) 'GM-gene-deletor': fused *loxP-FRT* recognition sequences dramatically improve the efficiency of FLP or CRE recombinase on transgene excision from pollen and seed to tobacco plants. *Plant Biotechnol. J.* 5: 263-274.

Matsunaga, E., Sugita, K., Ebinuma, H. (2002). Asexual production of selectable marker-free transgenic woody plants, vegetatively propagated species. *Mol. Breed* 10: 95–106.

Mayo, H. (1980) *The Theory of Plant Breeding. Oxford*: Clarendon Press.

Meyer, P., Saedler, H. (1996) Homology dependent gene silencing in plants. *Ann. Rev. Plant Physiol. Plant Mol. Biol.* 47: 23-48.

Miki, B., McHugh, S. (2004) Selectable marker genes in transgenic plants: applications, alternatives and biosafety. *J. Biotechnol.* 107: 193-232.

Mlynárová, L., Conner, A.J., Nap, J.-P. (2006) Directed microspore-specific recombination of transgenic alleles to prevent pollen-mediated transmission of transgenes. *Plant Biotechnol. J.* 4: 445-452.

Naiche, L.A., Papaioannou, V.E. (2007) Cre activity causes widespread apoptosis and lethal anemia during embryonic development. *Genesis* 45: 768-775.

Olivares, E.C., Hollis, R.P., Calos, M.P. (2001) Phage R4 integrase mediates site-specific integration in human cells. *Gene* 278: 167-176.

Onouchi, H., Yokoi, K., Machida, C., Matsuzaki, H., Oshima, Y., Matsuoka, K., Nakamura, K., Machida, Y. (1991) Operation of an efficient site-specific recombination system of Zygosaccharomyces rouxii in Tobacco cells. *Nucleic. Acids Res.* 19: 6373-6378.

Ow, D.W. (2002) Recombinase-directed plant transformation for the post-genomic era. *Plant Mol. Biol.* 48: 183-200.

Petolino, J.F., Worden, A., Curlee, K., Connell, J., Moynahan, T.L.S., Larsen, C., Russell, S. (2010) Zinc finger nuclease-mediated transgene deletion. *Plant Mol. Biol.* 73: 617-628.

Poehlman, J.M., Sleper, D.A. (1995) The genetic basis of plant breeding. In: Breeding field crops, 4th edition. Iowa State University Press, Ames, Iowa, pp 19-82.

Porteus, M.H., Carroll, D. (2005) Gene targeting using zinc finger nucleases. *Nat. Biotechnol.* 23: 967-973.

Prakash, N.S., Bhojaraja, R., Shivbachan, S.K., Priya, G.G.H., Nagraj, T.K., Prasad, V., Babu, V.S., Jayaprakash, T.L., Dasgupta, S., Spencer, T.M., Boddupalli, R.S. (2009) Marker-free transgenic corn plant production through co-bombardment. *Plant Cell Rep.* 28: 1655-1668.

Puchta, H. (2000) Removing selectable marker genes: taking the shortcut. *Trends Plant. Sci.* 5: 273-274.

Puchta, H., Hohn, B. (1996) From centiMorgans to basepairs: homologous recombination in plants. *Trends Plant Sci.* 1: 340-348.

Sadowski, P.D.E. (1993) Site-specific genetic recombination: hops, flips, and flops. *FASEB J.* 7: 760-767.

Sasaki, A., Ashikari, M., Ueguchi-Tanaka, M., Itoh, H., Nishimura, A., Swapan, D., Ishiyama, K., Saito, T., Kobayashi, M., Khush, G.S., Kitano, H., Matsuoka, M. (2002) A mutant gibberellin-synthesis gene in rice. *Nature* 416: 701-702.

Sauer, B. (1987) Functional expression of the *cre-lox* site-specific recombination system in the yeast *Saccharomyces cerevisiae. Mol. Cell Biol.* 7: 2087-2096.

Sauer, B., Henderson, N. (1988) Site-specific DNA recombination in mammalian cells by the Cre recombinase of bacteriophage P1. *Proc. Natl. Acad. Sci. USA* 93: 1156-1160.

Sauer, B., Henderson, N. (1989) Cre-stimulated recombination at *loxP*-containing DNA sequences placed into the mammalian genome. *Nucleic. Acids. Res.* 17: 147-161.

Schwikardi, M., Dröge, P. (2000) Site-specific recombination in mammalian cells catalyzed by γδ resolvase mutants: implications for the topology of episomal DNA. *FEBS Lett* 471: 147-150.

Smith, M.C., Thorpe, H.M. (2002) Diversity in the serine recombinases. *Mol. Microbiol.* 44: 299-307.

Sreekala, C., Wu, L., Gu, K., Wang, D., Tian, D., Yin, Z. (2005) Excision of a selectable marker in transgenic rice (*Oryza sativa* L.) using a chemically regulated Cre/loxP system. *Plant Cell Rep.* 24: 86-94.

Srivastava, V., Anderson, O.D., Ow, D.W. (1999) Single-copy transgenic wheat generated through the resolution of complex integration patterns. *Proc. Natl. Acad. Sci. USA* 96: 11117-11121.

Srivastava, V., Ow, D.W. (2001) Single-copy primary transformants of maize obtained through the co-introduction of a recombinase-expressing construct. *Plant Mol. Biol.* 46: 561–566.

Srivastava, V., Ow, D.W. (2004) Marker-free site-specific gene integration in plants. *Trends Biotechnol.* 22: 627-629.

Stark, W.M., Boocock, M.R., Sherratt, D.J. (1992) Catalysis by site-sepcific recombinases. *Trends Genet.* 8: 432-439.

Sternberg, N., Hamilton, D. (1981) Bacteriophage P1 site-specific recombination- I. Recombination between *lox*P sites. *J. Mol. Biol.* 150: 467-486.

Stoll, S., Ginsburg, D.S., Calos, M.P. (2002) Phage TP901-1 site-specific integrase functions in human cells. *J. Bacteriol.* 184: 3657–3663.

Takahashi, T., Naito, S., Komeda, Y. (1992) Isolation and Analysis of the Expression of Two Genes for the 81-Kilodalton Heat-Shock Proteins from *Arabidopsis*. *Plant Physiol.* 99: 383-390.

Thomason, L.C., Calendar, R., Ow, D.W. (2001) Gene insertion and replacement in *Schizosaccharomyces pombe* mediated by the *Streptomyces* bacteriophage phiC31 site-specific recombination system. *Mol. Gen. Genet.* 265: 1031-1038.

Thompson, C.J., Movva, N.R., Tizard, R., Crameri, R., Davies, J.E., Lauwereys, M., Botterman, J. (1987) Characterization of the herbicide-resistance gene *bar* from *Streptomyces hygroscopicus*. *EMBO J.* 6: 2519-2523.

Thomson, J.G., Chan, R., Thilmony, R., Yau, Y.-Y., Ow, D.W. (2010a) PhiC31 recombination system demonstrates heritable germinal transmission of site-specific excision from the *Arabidopsis* genome. *BMC Biotechnology* 10:17 (on-line).

Thomson, J.G., Ow, D.W. (2006) Site-specific recombination systems for the genetic manipulation of eukaryotic genomes. *Genesis* 44: 465-476.

Thomson, J.G., Yau, Y.-Y., Blanvillain, R., Chiniquy, D., Thilmony, R. Ow, D.W. (2009) ParA resolvase catalyzes site-specific excision of DNA from the *Arabidopsis* genome. *Transgenic Res.* 18: 237–248.

Thomson, J.G., Yau, Y. -Y, Chan, R., Thilmony, R. (2010b) Development of site-specific recombinase technology for precise crop plant genome modification. (*IAPB conference abstract, 2010*)

Thorpe, H.M., Wilson, S.E., Smith, M.C. (2000) Control of directionality in the site-specific recombination system of the Streptomyces phage phiC31. *Mol. Microbiol.* 38: 232-41.

Tobias, D.J., Manoharan, M., Pritsch, C., Dahleen, L.S. (2007) Co-bombardment, integration and expression of rice chitinase and thaumatin-like protein genes in barley (*Hordeum vulgare* cv. Conlon). *Plant Cell Rep.* 26: 631-639.

Verweire, D., Verleyen, K., De Buck, S., Claeys, M., Angenon, G. (2007) Marker-free transgenic plants through genetically programmed auto-excision. *Plant Physiol.* 145: 1220–1231.

Wang, M.-B., Waterhouse, P.M. (2000) High-efficiency silencing of a β-glucuronidase gene in rice is correlated with repetitive transgene structure but is independent of DNA methylation. *Plant Mol. Biol.* 43: 67–82.

Wang, Y., Chen, B., Hu, Y., Li, J., Lin, Z. (2005) Inducible excision of selectable marker gene from transgenic plants by the cre/lox site-specific recombination system. *Transgenic Res.* 14: 605-614.

Warwick, S.I., Simard, M.J., Légère, A., Beckie, H.J., Braun, L., Zhu, B., Mason, P., Séguin-Swartz, G., Stewart, C.N. (2003) Hybridization between transgenic *Brassica napus* L. and its wild relatives: *Brassica rapa* L., *Raphanus raphanistrum* L., *Sinapis arvensis* L., and *Erucastrum gallicum* (Willd.) O.E, Schulz. *Theor. Appl. Genet.* 107: 528-539.

Woo, H.J., Cho, H.S., Lim, S.H., Shin, K.S., Lee, S.M., Lee, K.J., Kim, D.H., Cho, Y.G. (2009) Auto-excision of selectable marker genes from transgenic tobacco via a stress inducible FLP/FRT site-specific recombination system. *Transgenic Res.* 18: 455-465.

Wulff, B.B.H., Kruijt, M., Collins, P.L., Thomas, C.M., Ludwig, A.A., De Wit, P.J.G.M., Jones, J.D.G. (2004) Gene shuffling-generated and natural variants of the tomato resistance gene *Cf-9* exhibit different auto-necrosis-inducing activities in *Nicotiana* species. *Plant J.* 40: 942-956.

Yau, Y.-Y., Wang, Y.J., Thomson, J.G., Ow, D.W. (2010) Method for Bxb1-mediated site-specific integration *in planta*. Ed. James A. Birchler. In: *Plant Chromosome Engineering: Method in Molecular Biology*. Humana Press, Clifton, New Jersey).

Zhang, W., Subbarao, S., Addae, P., Shen, A., Armstrong, C., Peschke, V., Gilbertson, L. (2003) Cre/lox-mediated marker gene excision in transgenic maize (*Zea mays* L.) plants. *Theor. Appl. Genet.* 107: 1157-1168.

Zhang, Y., Li, H., Ouyang, B., Lu, Y., Ye, Z. (2006) Chemical-induced autoexcision of selectable markers in elite tomato plants transformed with a gene conferring resistance to lepidopteran insects. *Biotechnol. Lett.* 28: 1247-1253.

Zhang, Y., Liu, H., Zhang, J.-T., Li, Y., Zhang, H. (2009) Generation of selectable marker-free transgenic tomato resistant to drought, cold and oxidative stress using the Cre/*loxP* DNA excision system. *Transgenic Res.* 18: 607-619.

Zubko, E., Scutt, C., Meyer, P. (2000) Intrachromosomal recombination between attP regions as a tool to remove selectable marker genes from tobacco transgenes. *Nat. Biotechnol.* 18: 442-445.

Zuo, J., Niu, Q.-W., Møller, S.G., Chua, N.-H. (2000) An estrogen receptor-based transactivator XVE mediates highly inducible gene expression in transgenic plants. *Plant. J.* 24: 265-273.

Zuo, J., Niu, Q.-W., Møller, S.G., Chua, N.-H. (2001) Chemical-regulated, site-specific DNA excision in transgenic plants. *Nat. Biotechnol.* 19: 157–161.

In: Transgenic Plants
Editors: Shen Y. Zhu and Jiang L.Hu
ISBN 978-1-62257-245-8

Chapter 7

AMPLIFICATION OF SMALL INTERFERING RNAS IN TRANSGENIC PLANTS

Hiroaki Kodama*, Hazuki Iwasa, Sayaka Hirai and Shin-Ichiro Oka
Graduate School of Horticulture, Chiba University, Matsudo, Japan

ABSTRACT

RNA interference (RNAi) is a rapidly popularized technology which contributes to progress in basic plant sciences and applied plant biotechnology. Most plant RNAi vectors transcribe an inverted repeat sequence consisting of target sequences. The resultant hairpin RNAs are processed into 21~25-nucleotide-long small interfering RNAs (siRNAs). These primary siRNAs are incorporated into the ARGONAUTE1-containing protein complexes, and then guide the sequence-specific cleavage of mRNAs. Primary siRNAs often trigger the synthesis of secondary siRNAs. In this amplification process, annealing of primary siRNAs to target mRNAs is followed by recruitment of RNA-DEPENDENT RNA POLYMERASE6 (RDR6), and then complementary RNA molecules against the target mRNAs are synthesized. The newly synthesized dsRNAs are processed into secondary siRNAs by redundant activities of DICER-LIKE4 and DICER-LIKE2. Since the target region of secondary siRNAs spreads into regions outside the primary target site,

* Corresponding authoer: E-mail: kodama@faculty.chiba-u.jp.

this phenomenon is called transitivity. Spreading of RNAi target via transitivity deteriorates specificity of RNAi. Therefore, elucidation of mechanistic issues of transitivity is important for researchers who use RNAi plants. Transitivity is frequently observed in transgenic plants expressing target transgene sequences under the presence of primary siRNAs, and there is only one report showing transitivity along the endogenous mRNA templates. It remains unclear why most endogenous transcripts seem to be an inefficient template for RDR6 during transitivity, but the limited transitivity along the endogenous transcripts apparently contributes to the maintenance of specificity of RNAi. In this review, we focus on transitivity observed in transgenic plants.

INTRODUCTION

To elucidate the function of unidentified genes, we often introduce genes of interest into plants, and investigate the effects of introduced genes on the expression of other related genes and also on the phenotype of resulting transgenic plants. Such approaches using transgenic plants bring us a wealth of information about the function of target genes. Although the transgene is expressed under the control of a linked promoter, its transcriptional activity often significantly differs among the transgenic plant lines. This epigenetic modulation of transcriptional activity is apparently ascribed to differences in the insertion site of transgenes into the plant genome and also to variations in the organization pattern of the transgene locus such as repetitiveness of transgene copies (Hamada and Kodama, 2006). By contrast, down-regulation of the target gene has been achieved by using several different methods. The antisense strategy was first used by Izant and Weintraub (1984) in order to reduce eukaryotic gene expression. Antisense RNAs are expected to form RNA duplexes with the sense mRNA molecules, which is followed by inhibition of translation or degradation of double-stranded RNA (dsRNA). Here we designated silencing by antisense technology as antisense post-transcriptional gene silencing (AS-PTGS). AS-PTGS had been used as the main method of inactivation of target genes in the 1990s (Dougherty and Parks, 1995). Around 1990, an unexpected phenotype of transgenic plants was reported (Matzke et al., 1989; Hobbs et al., 1990; Napoli et al., 1990; van der Krol et al., 1990). For example, a gene encoding petunia chalcone synthase (*CHS*), one of the key enzymes of anthocyanin synthesis, was introduced into non-white, pigmented flowered petunia cultivars. Several transgenic lines, but not all, showed entirely white flowers. In this case, the sense *CHS* transgene sequences caused a knockdown phenotype of endogenous *CHS* genes (Jorgensen, 1995).

This silencing phenomenon triggered by the sense transgene is called cosuppression, post-transcriptional gene silencing (PTGS) and/or sense post-transcriptional gene silencing (S-PTGS). S-PTGS can be observed in plants transformed with reporter genes including β-glucuronidase (*GUS*) gene and green fluorescence protein (*GFP*) gene, where inactivation of transgenes can occur among two or more transgene copies (Elmayan and Vaucheret, 1996; Boutla et al., 2002). We often observed more enhanced inhibition of target gene expression in S-PTGS plants than in AS-PTGS plants (Hamada et al., 1996; Davuluri et al., 2004; Tomita et al., 2004), and S-PTGS had frequently been used as a silencing technology until around 2000 in the field of plant biology. Then another technology, RNA interference (RNAi), replaced AS- and S-PTGS methods because of its high efficacy and superiority in silencing (e.g., Wang and Wagner, 2003). The molecular basis of S-PTGS is now considered to be similar to that of RNAi. This sequence-specific inhibition of target gene expression is mediated by a unique class of RNA, a 21~25-nucleotide (nt)-long small RNA species. Elucidation of the action mechanisms of small RNAs reveals diverse pathways including a dynamic amplification process of small RNAs. Lipardi et al. (2001) describe it as "nature's own PCR machine". Here we mainly focus on amplification of small RNAs which can potentially affect the phenotype of transgenic plants.

1. Outline of RNAi and Related Silencing Pathways

RNAi was discovered in a worm, *Caenorhabditis elegans*, by Fire, Mello and their colleague. Simultaneous introduction of sense and antisense RNA molecules provokes sequence-specific gene inactivation (Fire et al., 1998). Now, RNAi has become a potent tool to elucidate gene functions in a genome-wide manner (e.g., Kamath et al., 2003; Reddien et al., 2005). RNAi is mediated by about 21~25-nt long small interfering RNAs (siRNAs). Interestingly, siRNAs were discovered from plants showing several silenced phenotypes including S-PTGS (Hamilton and Baulcombe, 1999). A single-stranded siRNA is associated with an ARGONAUTE protein (AGO), a component of RNA-induced silencing complex (RISC; Hammond et al., 2000). mRNAs are cleaved at 10 nt from the 5′ end of a 21-nt siRNA by the "slicer" (Song et al., 2004). In *Arabidopsis thaliana*, ten predicted *AGO* family members (*AGO1~AGO10*) have been identified, and AGO1 is the main slicer in S-PTGS and RNAi (Baumberger and Baulcombe, 2005; Brodersen and Voinnet, 2006).

In mammalian cells and worms, RNAi can be triggered by direct introduction of siRNA molecules into cells. In plants, most RNAi have been established by introducing a transcriptional unit producing long dsRNAs because plant cells are not amenable to direct transfer of nucleic acids. Plant RNAi vectors contain a transcriptional unit harboring an inverted repeat sequence, and produce a hairpin RNA (Chuang and Meyerowitz, 2000; Smith et al., 2000). The stem region of hairpin RNAs is recognized as a dsRNA and then processed into siRNAs by an endonucleolytic enzyme, Dicer (in plants, usually called DICER-LIKE protein, DCL) (Baulcombe, 2004). In *Arabidopsis*, four DCL genes (*DCL1~DCL4*) have been identified. Until recently, no mutant defective in RNAi had been isolated. Since a high level of dsRNAs may activate different RNAi-related pathways, the true genetic requirement of RNAi may be bypassed. By using a phloem-restricted, moderately expressed SUC promoter, Dunoyer et al. (2007) indicate that DCL1 excises the stem-loop portion from hairpin RNAs and 21- and 24-nt siRNAs are processed from the stem by DCL4 and DCL3, respectively. In the absence of DCL4, the stem portion was processed into 22-nt siRNAs by DCL2. When hairpin RNAs are frequently transcribed by a strong promoter such as the cauliflower mosaic virus 35S promoter, both DCL2 and DCL4 are likely to be involved in the 21-nt siRNA production (Fusaro et al., 2006; Mlotshwa et al., 2008). The majority of small RNAs associated with the main slicer, AGO1, are 21 nt in length (Mi et al., 2008). The 24-nt siRNAs are incorporated into AGO4 and involved in DNA methylation and transcriptional gene silencing (so-called TGS) at specific genomic loci (Matzke et al., 2007) (Figure 1a).

siRNAs are also a key substance in S-PTGS, but nascent transcripts from the sense transgene cannot serve as a substrate for Dicer. Thus a step in which sense transcripts are converted into dsRNAs is required. RNA-DEPENDENT RNA POLYMERASE (RDR) is responsible for production of complementary RNAs (cRNAs). In *Arabidopsis*, there are six *RDR* genes (*RDR1~RDR6*) (Wassenegger and Krczal, 2006). The *sde1/sgs2* mutants had been isolated as plants defective in S-PTGS (Dalmay et al., 2000; Mourrain et al., 2000), and their loci were identified as a loss-of-function *RDR6* allele (Wassenegger and Krczal, 2006). *RDR6* is a negligible component in the dsRNA-triggered RNAi (Figure 1b). An endogenous small RNA class, microRNA (miRNA), is involved in the developmental regulation and stress response (Jones-Rhoades et al., 2006; Mallory and Vaucheret, 2006). Both sense and antisense sequences required for dsRNA formation are encoded on the genome as a *bona-fide* regulatory locus. After transcription, imperfect stem-loop transcripts (pri-miRNAs) are processed by DCL1 and then miRNAs are produced (Kurihara and Watanabe, 2004). Most miRNAs are also incorporated into AGO1 (Mi et al., 2008) (Figure 1c).

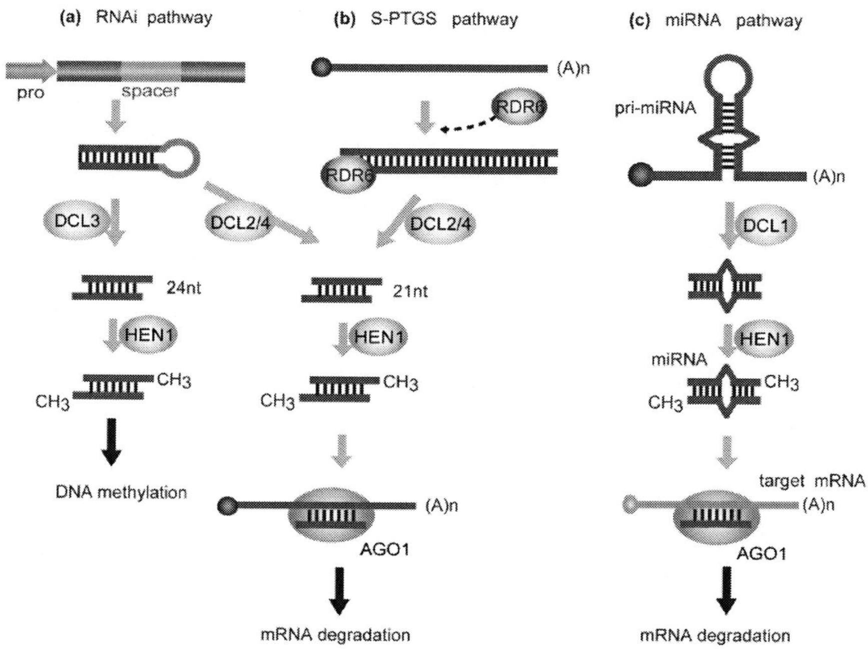

Figure 1. Representative silencing pathways mediated by siRNAs and miRNAs. (a) RNAi pathway. An RNAi construct produces hairpin RNAs. DCL3 processes dsRNAs into 24-nt-long siRNAs which direct cytosine methylation on the homologous genomic loci. DCL4 produces 21-nt-long siRNAs. DCL2 is also involved in this process. HEN1 mediates methylation at the 3′ termini of the siRNAs (Yu et al., 2005). The siRNA incorporated into AGO1 guides the endonucleolytic cleavage of homologous RNAs. pro, promoter. (b) S-PTGS pathway. Transcripts from the sense transgene are recognized as "aberrant" RNAs and then RDR6 is recruited. RDR6 synthesizes cRNA molecules. These dsRNAs are then processed into 21-nt siRNAs by DCL4 and DCL2. (c) miRNA pathway. miRNAs are transcribed as long primary transcripts (pri-miRNAs) and processed into miRNAs by DCL1. A few miRNAs act as an inhibitor of translation (Chen, 2004), but the majority of miRNAs are considered to guide mRNA cleavage.

2. Phenotypic Variation in Plants Showing RNA Silencing

siRNAs diagnostic for S-PTGS/RNAi are also found in some AS-PTGS plants, suggesting that AS-PTGS, S-PTGS and RNAi partly share a common mechanism (Di Serio et al., 2001). Inhibitory phenomena mediated by small RNAs are generically called RNA silencing (Baulcombe, 2004). When an endogenous gene was inactivated by using different silencing strategies, phenotypic variations were observed among the transgenic plants showing RNA silencing. Jorgensen and colleague scored many transgenic plants in which expression of the *CHS* gene was modulated by AS-PTGS, S-PTGS and also by transgene loci with inverted-repeat structures. Since such inverted-repeat structured transgene loci were found in the petunias transformed with a sense construct, we here designate the silencing via these loci as the inverted repeat-structured, sense-PTGS (IR-S-PTGS) in order to discriminate from the silencing via RNAi. These silenced loci produced several different pigmentation patterns in flowers (Jorgensen et al., 1996; Que and Jorgensen, 1998); S-PTGS produced a morphology-based flower pigmentation pattern, AS-PTGS exhibited weakened pigmentation and IR-S-PTGS produced "complex patterns" with sometimes disordered patterns. The initiation mechanism for S-PTGS should include the interaction between transcripts from the transgene and endogenous genes, which results in the developmental or/and tissue-specific control of S-PTGS.

3. Caveat to siRNA-Mediated Silencing

siRNA molecules play a pivotal role in the determination of sequence specificity in RNA silencing, but it was reported that siRNAs often down-regulated mRNA species other than the intended target mRNA. This phenomenon is so-called "off-target effect" (Jackson et al., 2003). Qiu and Lane (2007) investigated the number of mismatches between siRNA and mRNA that can be tolerated for silencing of the target gene. Mismatches in the siRNA-mRNA pair of 3 to 4 nt were usually permissive for silencing activity in *Drosophila* and mammalian cells (Elbashir et al., 2001; Jackson et al., 2003; Saxena et al., 2003). The higher permissiveness of mismatches apparently increases the probability of off-target silencing. Simulation in *Schizo-saccharomyces pombe* indicates that more than 6 nt mismatches would turn

RNAi from defensive to self-destructive. Thus, the upper limit of permissive mismatches is no more than 6 nt (Qiu and Lane, 2007). Unlike flies, worms and mammals, mismatches in a single siRNA cannot be controlled in most RNAi plants. The plant RNAi is guided by many siRNA species processed from a relatively long stem region of hairpin RNAs (usually 100~500 nt in length).

Therefore, target mRNAs may be more preferentially degraded than other unintended mRNAs, but simultaneously opportunity of off-target silencing increases in the RNAi plants. Xu et al. (2006) developed a Web-based computational tool to identify potential off-targets. Although a few mismatches in the siRNA-mRNA pair is expected to be permissive for establishment of RNA silencing, a varied reduction of expression levels of the potential off-targets has been observed in plants. In fact, sequence identity with at least 22-nt continuous stretches is required for off-target silencing in *Arabidopsis* plants. Another caveat about the specificity of siRNA-mediated RNA silencing is transitivity, a phenomenon in which primary siRNAs trigger transformation of target mRNAs into dsRNAs and new siRNAs (secondary siRNAs) are produced. RDR6 is recruited during transitivity for production of dsRNAs. The absence of the RDR family in the *Drosophila* and human genomes is in agreement with a finding that plants and worms show transitivity while human cells and *Drosophila* do not (Sijen et al., 2001; Vaistij et al., 2002; Zamore, 2002). Secondary as well as primary siRNAs can mediate mRNA cleavage (Van Houdt et al., 2003; Bleys et al., 2006a, b), and in some cases, generation of secondary siRNAs is associated with more enhanced silencing (Shimamura et al., 2007). Thus amplification of siRNAs during transitivity leads to accelerated mRNA cleavage and also gives rise to increased chances of off-target RNA silencing.

4. Generation of Secondary siRNAs in Transgenic Plants

Production of secondary siRNAs has been observed when the primary siRNA-producing locus exists with a transgene locus producing sense transcripts harboring the target sequences of primary siRNAs. For example, secondary siRNA production was observed when transgenic plants expressing a sense transgene were infected with plant viruses carrying parts of the same transgene (Vaistij et al., 2002). Infection of plant viruses induces generation of siRNAs against the viral sequences, and this phenomenon has been called as virus-induced gene silencing (VIGS) (Baulcombe, 1999; Waterhouse et al., 1999). Alternatively, secondary siRNAs are generated in crossbred plants between the sense transgenic plants and RNAi-plants carrying a partial region of the same transgene sequence in an inverted repeat manner (Haque et al., 2007; Shimamura et al., 2007). Secondary siRNAs were also generated in sense transgenic plants when RNAi constructs harboring part of the sense transgene sequences were introduced by agroinfiltration (Himber et al., 2003; Oka et al., 2008), and when dsRNAs were introduced by a biolistic method (Klahre et al., 2002). As expected from these observations, S-PTGS fulfills the conditions required for secondary siRNA production, i.e. co-existence of sense transcripts and corresponding siRNAs. Therefore, S-PTGS must conclude an siRNA amplification process (Qiu and Lane, 2007). miRNAs-induced transitivity has also been reported. To quantify miRNA-mediated suppression, Parizotto et al. (2004) prepared a *GFP*-sensor construct which has the authentic target sequence of miR171 in its 3′ untranslated region. The resulting transcripts were cleaved by miR171 and then *GFP* secondary siRNAs were produced (Moissiard et al., 2008). In addition to the cell-autonomous production mentioned above, secondary siRNAs are also produced in a non-cell-autonomous manner. Spreading of RNA silencing from local tissues into the whole plant is associated with the long-range movement of silencing signals (Voinnet et al., 1998). This systemic gene silencing requires a trancriptionally active transgene in the recipient tissues (Jorgensen et al., 1998). Grafting of non-silenced scions onto rootstocks showing RNA silencing leads to RNA silencing in the scion. When rootstocks were prepared from the RNAi plants carrying the partial fragment of the target gene, transitivity was observed in the systemically silenced scion (Brosnan et al., 2007; Haque et al., 2007; Shimamura et al., 2007).

5. RDR6-Dependent Transitive RNA Silencing

In *C. elegans*, siRNAs guide mRNA cleavage, and they simultaneously serve as primers for RDR, leading RNAi to spread from the primary target region into adjacent upstream regions of the target gene (Sijen et al., 2001). In plants, generation of secondary siRNAs is a more complex process than in worms. Secondary siRNAs are synthesized from the 5′ upstream region and/or 3′ downstream region adjacent to the inducer sequences. The 5′ and 3′ spreading of siRNA generation is not exclusive (Voinnet et al., 1998; Klahre et al., 2002; Vaistij et al., 2002), but unidirectional transitivity is also often observed. In *Arabidopsis* and tobacco plants, RDR6 is responsible for production of dsRNA in transitivity (Vaistij et al., 2002; Himber et al., 2003; Schwach et al., 2005; Oka et al., 2008). The newly synthesized dsRNAs are then processed by the redundant functions of DCL2 ad DCL4, and most secondary siRNAs are 21 nt in length (Mlotshwa et al., 2008; Moissiard et al., 2008). Two possible mechanisms for 5′ secondary siRNA production have been proposed. After cleavage of target mRNAs by primary siRNAs, the resulting 3′ truncated, poly(A)-less RNAs are recognized as a substrate for RDR6. Alternatively, primary siRNAs act as primers for RDR6. Moissiard et al. (2008) addressed the 5′ secondary siRNA production along the *GFP*-sensor construct. When three mismatches corresponding to the 3'-end of miR171 were introduced into the *GFP*-sensor construct, miR171 guided the cleavage of this modified *GFP*-sensor transcripts. However, the 5′ secondary *GFP* siRNAs were below detection level. This result strongly suggests that primary siRNAs act as primers of RDR6 during transitivity into the 5′ upstream region. In the case of preferential production of 3′ secondary siRNAs (Petersen and Albrechtsen, 2005; Haque et al., 2007; Shimamura et al., 2007), two possible synthetic pathways have been proposed. The uncapped, 5′ truncated mRNA fragments generated by primary-siRNA mediated cleavage are recognized by RDR6 and the 3′ truncated ones are discarded. Alternatively, the processibility of RDR6 is limited and cRNA molecules corresponding to the 5′ upstream region are hardly synthesized on the template which has a long gap between the binding site of primary siRNAs and the 3′ end of templates.

In this respect, Oka et al. (2008) investigated 3′ unidirectional production of secondary siRNAs by using several RNAi constructs that harbored target regions spanning the entire sense transcript. These RNAi constructs were infiltrated into the leaves of transgenic plants expressing the target gene. The results showed that the 5′ secondary siRNAs were hardly produced irrespective of the location of inducer sequences on the sense transcript, suggesting that the former hypothesis, namely RDR6 is preferentially recruited on the 5′ truncated, uncapped RNA fragment. Gazzani et al. (2004) showed that uncapped RNAs are a preferred template for RDR6. Such uncapped RNAs are rapidly degraded by a 5′-3′ exoribonuclease (XRN4) in normal conditions. When XRN4 is deficient, uncapped transcripts produced irregularly from a highly active transgene locus escape being degraded. These transcripts are recognized as a template of RDR6 and siRNAs are produced.

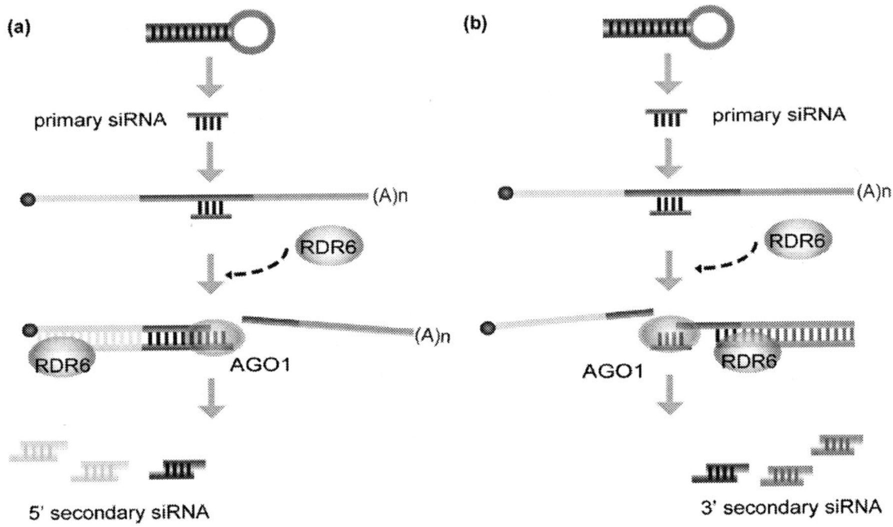

Figure 2. Two modes of transitivity. (a) 5′ unidirectional transitivity. Primary siRNAs processed from hairpin RNAs act as primers of RDR6. RDR6 synthesizes cRNAs by using poly(A)-less, 3′ truncated transcripts as a template. (b) 3′ unidirectional transitivity. After cleavage of target mRNAs, the cap-less, 5′ truncated transcripts are recognized by RDR6. Note that these two pathways are not always exclusive.

Therefore, the transgene locus was converted into the S-PTGS locus in XRN4-deficient cells. Taken together, transitivity seems to occur as follows; target mRNAs are cleaved under the guidance of primary siRNAs, which is followed by primer-dependent cRNA synthesis on 3′ truncated RNA fragments (Figure 2a) and primer-independent cRNA synthesis on 5′ truncated, uncapped RNA fragments (Figure 2b). RDR6 is responsible for the cRNA synthesis corresponding to both the 5′ upstream and 3′ downstream regions. Both primer-dependent and -independent RDR activities were observed in wheat germ extracts (Tang et al., 2003). It is unclear how RDR6 chooses its templates in the case of unidirectional transitive RNA silencing. The finely regulated recruitment of RDR6 was recently found during the biogenesis of *trans*-acting siRNAs (tasiRNAs). Several endogenous, non-coding transcripts (pri-tasiRNAs) can serve as a template for RDR6 (Peragine et al., 2004). The pri-tasiRNAs harbor the target sequences against specific miRNA. After miRNA-mediated cleavage, pri-tasiRNAs are converted into dsRNA by RDR6, and then tasiRNAs are produced by DCL4 (Vazquez et al., 2004; Allen et al., 2005; Yoshikawa et al., 2005). *TAS3*, one of the pri-tasiRNAs, have two target sites of miR390, and Axtell et al. (2006) proposed a two-hit trigger model, in which transcripts with two or more small RNA target sites become a template for RDR6. These sites function specifically with AGO7 in order to recruit RDR6 (Montgomery et al., 2008). AGO7 functions as a miR390-specific slicer (Montgomery et al., 2008). Production of other tasiRNAs from the *TAS1*, *TAS2* and *TAS4* pri-tasiRNAs are independent of AGO7. How RDR6 is recruited on the *TAS1*, *TAS2* and *TAS4* templates is still unclear.

6. Transgene Transcripts: Preferable Templates for RDR6

Production of tasiRNAs initiates with cleavage of pri-tasiRNAs under guidance by miR173 (*TAS1* and *TAS2*), miR390 (*TAS3*) and miR828 (*TAS4*), but the majority of other plant miRNA targets do not become a template of RDR6. In fact, miR171 can induce transitivity along the *GFP*-sensor transcript (Moissiard et al., 2008), but phased siRNAs from endogenous targets of miR171 were not detected in the genome-wide analysis of siRNAs produced via the RDR6/DCL4 pathway (Howell et al., 2007). Similarly, endogenous transcripts seldom serve as the RDR6 template during transitivity in the transgenic RNAi plants and VIGS-induced plants. One exceptional case was reported in the S-PTGS with the glucanase gene (Sanders et al., 2002). The glucanase transgene caused cosuppression of the endogenous glucanase genes which showed various degrees of homology with the transgene, and siRNAs corresponding to these silenced endogenous genes were detected. However, most other studies showed that secondary siRNA production preferentially occurs on the transgene template and not on the endogenous ones (Vaistij et al., 2002; Kościańska et al., 2005; Miki et al., 2005; Petersen and Albrechtsen 2005; Bleys et al., 2006b; Shimamura et al., 2007). Therefore, endogenous transcripts are protected from transitivity by unidentified intrinsic features. At present, it is unclear why transgene transcripts are a preferable template of RDR6. The lack of amplification of secondary siRNAs along the endogenous mRNAs is apparently independent on the amount of target mRNAs because abundant mRNAs transcribed from the endogenous gene, such as the phytoene desaturase gene, the *rbcS* gene, and the magnesium chelatase subunit I gene did not serve as a template for RDR6 (Vaistij et al., 2002; Petersen and Albrechtsen, 2005). Wassenegger and Krczal (2006) suggest two possible mechanisms in which endogenous transcripts are protected from recognition of RDR6. One is the RNA surveillance mechanism associated with RNA processing. Most endogenous genes have introns, and their transcripts are processed by a spliceosome. Aberrant transcripts from the endogenous genes might be eliminated before translation. By contrast, most transgenes do not contain introns and the aberrant transcripts produced during highly active transcription are likely to be exported into the cytoplasm without the spliceosome-associated RNA surveillance. Another possible mechanism is the involvement of viral- and *Agrobacterium*-related regulatory sequences, such as viral promoters, nopaline synthase promoter and terminator sequences. These elements may be a cause of incomplete assembly of transcriptional complexes or changes of chromatin structures.

7. Possible Involvement of Nuclear RNA Silencing Pathway during Transitivity

Although production of secondary siRNAs depends on RDR6/DCL4 (and/or DCL2), recent analyses suggest involvement of the chromatin silencing pathway. This chromatin silencing seems to be mediated by 24-nt siRNAs possibly produced via combination of a large subunit of RNA polymerase IV (NRPD1a), RDR2, DCL3 and AGO4 (Matzke et al., 2007). The 24-nt siRNAs are presumed to direct DNA methylation, which are followed by histone modification and eventually lead to heterochromatinization (Xie et al., 2004). In addition, presence of 24-nt siRNAs is correlated with systemic RNA silencing (Hamilton et al., 2002). Thus, 24-nt siRNAs are considered to be involved in both nuclear silencing and systemic silencing pathways. Brosnan et al. (2007) grafted the scions prepared from mutants of *NRPD1a*, *RDR2*, *DCL3*, and *AGO4* onto the silenced rootstock.

All mutations compromised in the reception of silencing signals from the rootstock. RDR6 is required for generation of both 21- and 24-nt siRNAs. Since establishment of systemic gene silencing is associated with the generation of 3′ secondary siRNAs (Brosnan et al., 2007; Haque et al., 2007; Shimamura et al., 2007), the chromatin silencing pathway is likely involved in the generation of non-cell-autonomous secondary siRNAs. Interestingly, mutations of *NRPD1a* and the cytosine DNA methylase gene (*MET1*) caused weak defects in S-PTGS (Morel et al., 2000; Herr et al., 2005), implying that cell-autonomous secondary siRNA production during S-PTGS includes the nuclear chromatin silencing pathway.

8. Case Study: 24-nt siRNAs Produced from an Intron-Containing Transgene

As mentioned above, primary siRNAs processed from hairpin RNAs did not induce transitivity along the endogenous transcripts. One possible explanation is that endogenous nascent transcripts are discriminated by binding to the spliceosome. To test whether spliceosome-dependent processing is involved in discrimination of the endogenous transcripts, we expressed an intron-containing transgene in the RNAi and WT plants. We have investigated RNA silencing of tobacco endoplasmic reticulum ω-3 fatty acid desaturase gene (*NtFAD3*) (Hamada and Kodama, 2006). The *NtFAD3* RNAi plants (R11 line) were produced by introduction of an RNAi vector, pTF1AGS (Figure 3a). This plasmid contained an inverted repeat sequence corresponding to nucleotides 181-678 of the *NtFAD3* cDNA (GenBank acc. no. D26509; Tomita et al., 2004).

The 3′ secondary siRNA production can be observed in the crossbred plants between the R11 plant and the sense *NtFAD3* transgenic plant (Shimamura et al., 2007). Then, we examined whether or not the 3′ secondary siRNAs are produced by ectopic expression of the genomic fragment of the *NtFAD3* gene in the R11 leaves. The 1,512-bp-long *NtFAD3* genomic fragment corresponding to the *NtFAD3* cDNA from nucleotide position 339-1269 was inserted behind the cauliflower mosaic virus 35S promoter to generate a binary vector, pN3G-2/9. This plasmid expresses a genomic region from exon 2 to exon 9 of the *NtFAD3* gene. In addition, a corresponding cDNA fragment (nucleotide position 339-1269) was also cloned into a binary vector to produce a plasmid, pN3C-2/9 (Figure 3a). pN3G-2/9 and pN3C-2/9 were introduced into the leaves of R11 plants, WT plants and control transgenic plants (L1) by agroinfiltration. This control transgenic plant was produced by introduction of a sense construct (pBI-LUC) which harbors a firefly luciferase gene.

The L1 line was used as a representative one for successful overexpression of the luciferase gene. Transcripts from pN3G-2/9 and pN3C-2/9 could be detected in the WT plants but they were degraded by primary siRNAs produced from the R11 locus (Figure 3b). Intronic sequences were lacking in these transcripts when they were detected by Northern analysis and also by RT-PCR analysis (data not shown). siRNAs were detected by hybridization of DIG-labeled sense riboprobe covering the 3′ downstream *NtFAD3* region relative to the target sequences (D-probe; Figure 3a).

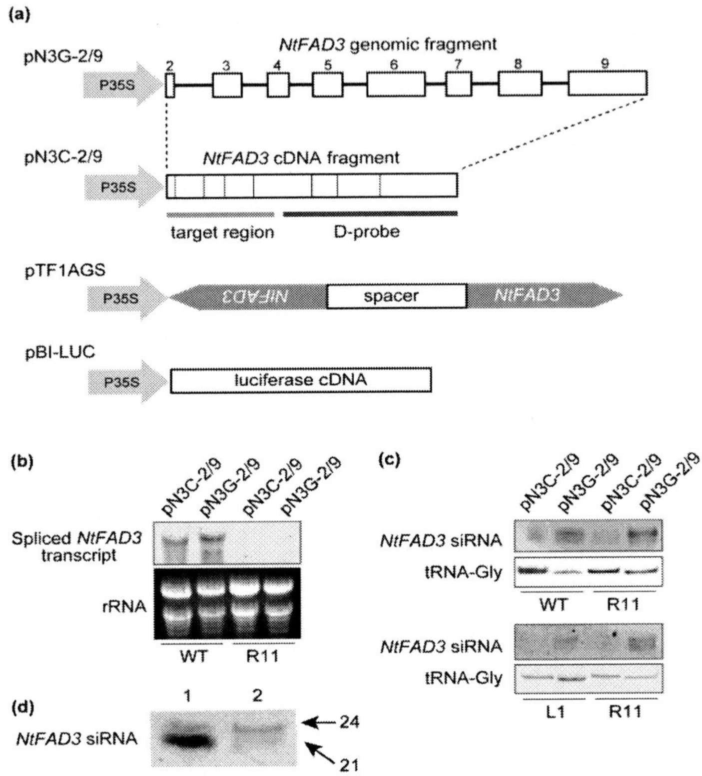

Figure 3. Preferential synthesis of 24-nt siRNAs by expression of an intron-containing transgene. (a) Constructs used. pN3G-2/9 harbors the *NtFAD3* genomic fragment (nucleotide positions 924 - 2435, GenBank acc. no. AB049576). The thick line shows introns, and white boxes show *NtFAD3* exons. pN3C-2/9 harbors a *NtFAD3* cDNA fragment (nucleotide positions 339-1269, Genbank acc. no. D26509) which corresponds to the genomic region used in construction of pN3G-2/9. The target region of an RNAi construct, pTF1AGS, is indicated. The region covered by a riboprobe (D-probe) is also shown. pTF1AGS is used for production of RNAi plants (R11 line) against the *NtFAD3* gene. pBI-LUC was introduced into tobacco plants to produce a control transgenic plant (L1 line). (b) Expression of pN3G-2/9 and pN3C-2/9 in infiltrated leaves. The transcript levels derived from these constructs were determined by Northern blot analysis. The equivalence of RNA loading among the lanes was demonstrated by ethidium bromide staining of rRNA bands. (c) Detection of the *NtFAD3* siRNAs in the infiltrated leaves prepared from WT, R11 and L1 plants. The *NtFAD3* siRNAs were detected by Northern blot hybridization with a sense D-probe. The blot was reprobed with a gene for tRNA-Gly(GCC) to verify that approximately the same amount of small RNA was loaded per lane (Oka et al., 2008). (d) *NtFAD3* siRNAs accumulated in the S-PTGS plants (Tomita et al., 2004; lane 1) and in the leaves agroinfiltrated with pN3G-2/9 (lane 2).

siRNAs hybridized with D-probes were preferentially detected in leaves of R11, L1 and also WT plants after infiltration with pN3G-2/9, but not with pN3C-2/9 (Figure 3c). Since predicted mature transcripts from these two binary vectors have the same sequence, siRNAs detected by a D-probe originated from the transgene transcripts themselves and did not result from transitivity. The majority of siRNAs accumulated in the S-PTGS plants against the *NtFAD3* gene were 21 nt in length (Tomita et al., 2004), but those produced by the transient expression of *NtFAD3* genomic sequences were mostly 24 nt in length (Figure 3d). The riboprobes against the intronic sequence did not give any hybridization signals (data not shown), indicating that most 24-nt *NtFAD3* siRNAs harbored exonic sequences. As discussed in the previous section, siRNAs involved in the nuclear RNA silencing pathway are mostly 24-nt long (Xie et al., 2004). Thus, ectopic expression of the partial *NtFAD3* genomic sequence may activate the nuclear silencing pathway, in which templates of RDR are seemingly the intron-removed transcripts. Thus it is probable that splicing-associated RNA surveillance cannot eliminate completely the aberrant RNA species (putative preferable templates of RDR6) during active transcription of the transgene.

CONCLUSION

Amplification of siRNAs in transgenic plants is an attractive phenomenon in which the silencing efficiency is expected to increase, but the frequency of silencing against unintended targets is also considered to increase. RDR6 recognizes the transgene-derived transcripts as its template during transitivity, and transcripts from the endogenous counterparts are apparently inefficient templates. Several RNAi-related factors may interact with RDR6 in the process of recognition of its template as was the case of synthesis of *trans*-acting siRNAs (Montgomery et al., 2008). In addition, strict discrimination of endogenous transcripts from those of the trangene suggest that the transgene has the intrinsic differences from other genomic sequences. For example, several silencing-related mutants were identified in a screen for changed expression of the reporter gene driven by the stress/defense-responsible promoters. The loci for these mutants are identified as loss-of-function *AGO4* and *ROS1* (a gene encoding a putative DNA repair protein involved in demethylation) (Gong et al., 2002; Agorio and Vera, 2007).

Therefore, these promoter-reporter transgene cassettes had been marked by the nuclear RNA silencing machinery. We should carefully evaluate the transgenic phenotype even though the transgene is apparently expressed in an expected manner.

REFERENCES

Agorio, A. & Vera, P. (2007). ARGONAUTE4 is required for resistance to *Pseudomonas syringae* in *Arabidopsis*. *Plant Cell*, *19*, 3778-3790.

Allen, E., Xie, Z., Gustafson, A. M. & Carrington, J. C. (2005). microRNA-directed phasing during *trans*-acting siRNA biogenesis in plants. *Cell*, *121*, 207-221.

Axtell, M., Jan, C., Rajagopalan, R. & Bartel, D. P. (2006). A two-hit trigger for siRNA biogenesis in plants. *Cell*, *127*, 565-577.

Baulcombe, D. C. (1999). Fast forward genetics based on virus-induced gene silencing. *Curr. Opin. Plant Biol.*, *2*, 109-113.

Baulcombe, D. (2004). RNA silencing in plants. *Nature*, *431*, 356-363.

Baumberger, N. & Baulcombe, D.C. (2005). *Arabidopsis* ARGONAUTE1 is an RNA slicer that selectively recruits microRNAs and short interfering RNAs. *Proc. Natl. Acad. Sci. USA*, *102*, 11928-11933.

Bleys, A., Van Houdt, H. & Depicker, A. (2006a). Down-regulation of endogenes mediated by a transitive silencing signal. *RNA*, *12*, 1633-1639.

Bleys, A., Vermeersch, L., Van Houdt, H. & Depicker, A. (2006b). The frequency and efficiency of endogene suppression by transitive silencing signals is influenced by the length of sequence homology. *Plant Physiol.*, *142*, 788-796.

Boutla, A., Kalantidis, K., Tavernarakis, N., Tsagris, M. & Tabler, M. (2002). Induction of RNA interference in *Caenorhabditis elegans* by RNAs derived from plants exhibiting post-transcriptional gene silencing. *Nucleic Acid Res.*, *30*, 1688-1694.

Brodersen, P. & Voinnet, O. (2006). The diversity of RNA silencing pathways in plants. *Trends Genet.*, *22*, 268-280.

Brosnan, C. A., Mitter, N., Christie, M., Smith, N. A, Waterhouse, P. M. & Carroll, B. J. (2007). Nuclear gene silencing directs reception of long-distance mRNA silencing in *Arabidopsis*. *Proc. Natl. Acad. Sci.* USA, *104*, 14741-14746.

Chen, X. (2004). A microRNA as a translational repressor of *APETALA2* in *Arabidopsis* flower development. *Science*, *303*, 2022-2025.

Chuang, C. F. & Meyerowitz, E. M. (2000). Specific and heritable genetic interference by double-stranded RNA in *Arabidopsis thaliana*. *Proc. Natl. Acad. Sci.*, USA, *97*,4985-4990.

Dalmay, T., Hamilton, A., Rudd, S., Angell, S. & Baulcombe, D. C. (2000). An RNA-dependent RNA polymerase gene in *Arabidopsis* is required for posttranscriptional gene silencing mediated by a transgene but not by a virus. *Cell*, *101*, 543-553.

Davuluri, G. R., Van Tuinen, A., Mustilli, A. C., Manfredonia, A., Newman, R., Burgess, D., Brummell, D. A., King, S. R., Palys, J., Uhlig, J., Pennings, H. M. J. & Bowler, C. (2004). Manipulation of DET1 expression in tomato results in photomorphogenic phenotypes caused by post-transcriptional gene silencing. *Plant J.*, *40*, 344-354.

Di Serio, F., Schöb, H., Iglesias, A., Tarina, C., Bouldoires, E. & Meins, F., Jr. (2001). Sense- and antisense-mediated gene silencing in tobacco is inhibited by the same viral suppressors and is associated with accumulation of small RNAs. Proc. *Natl. Acad. Sci.*, USA, 98, 6506-6510.

Dougherty, W. G. & Parks, T. D. (1995). Transgenes and gene suppression: telling us something new? *Curr. Opin. Cell Biol.*, *7*, 399-405.

Dunoyer, P., Himber, C., Ruiz-Ferrer, V., Alioua, A., & Voinnet, O. (2007) Intra- and intercellular RNA interference in *Arabidopsis thaliana* requires components of the microRNA and heterochromatic silencing pathways. *Nat. Genet.*, *39*, 848-856.

Elbashir, S. M., Martinez, J., Patkaniowska, A., Lendeckel, W. & Tuschl, T. (2001). Functional anatomy of siRNA for mediating efficient RNAi in *Drosophila melanogaster* embryo lysate. *EMBO J.*, *20*, 6877–6888.

Elmayan, T. & Vaucheret, H. (1996). Expression of single copies of a strongly expressed 35S transgene can be silenced post-transcriptionally. *Plant J.*, *9*, 787-797.

Fire, A., Xu, S., Montgomery, M. K., Kostas, S. A., Driver, S. E. & Mello, C. C. (1998). Potent and specific genetic interference by double-stranded RNA in *Caenorhabditis elegans*. *Nature*, *391*, 806-811.

Fusaro, A. F., Matthew, L., Smith, N. A., Curtin, S. J., Dedic-Hagan, J., Ellacott, G. A., Watson, J. M., Wang, M., Brosnan, C., Carroll, B. J. & Waterhouse, P. M. (2006). RNA interference-inducing hairpin RNAs in plants act through the viral defense pathway. *EMBO Rep.*, *7*, 1168-1175.

Gazzani, S., Lawrenson, T., Woodward, C., Headon, D. & Sablowski, R. (2004). A link between mRNA turnover and RNA interference in *Arabidopsis*. *Science*, *306*, 1046-1048.

Gong, Z., Morales-ruiz, T., Ariza, R. R., Roldan-Arjona, T., David, L. & Zhu, J. (2002). ROS1, a repressor of transcriptional gene silencing in *Arabidopsis*, encodes a DNA glycosylase/lyase. *Cell, 111*, 803-814.

Hamada, T. & Kodama, H. (2006). Phenotype of the transgene in plants: expression and silencing. In J. A. Teixeira da Silva (Eds.), *Floriculture, Ornamental and Plant Biotechnology:* Advances and Topical Issues (1st edition, vol II, pp. 98-107). London, UK: Global Science Books.

Hamada, T., Kodama, H., Nishimura, M. & Iba, K. (1996). Modification of fatty acid composition by over- and antisense-expression of a microsomal ω-3 fatty acid desaturase gene in transgenic tobacco. *Transgenic Res.*, *5*, 115-121.

Hamilton, A. J. & Baulcombe, D. C. (1999). A species of small antisense RNA in posttranscriptional gene silencing in plants. *Science*, *286*, 950-953.

Hamilton, A., Voinnet, O., Chappell, L. & Baulcombe, D. (2002). Two classes of short interfering RNA in RNA silencing. *EMBO J.*, *21*, 4671-4679.

Hammond, S. M., Bernstein, E., Beach, D. & Hannon, G. J. (2000). An RNA-directed nuclease mediates post-transcriptional gene silencing in *Drosophila* cells. *Nature*, *404*, 293-296.

Haque, A. K. M. N., Tanaka, Y., Sonoda, S. & Nishiguchi, M. (2007). Analysis of transitive RNA silencing after grafting in transgenic plants with the coat protein gene of *Sweet potato feathery mottle virus*. *Plant Mol. Biol.*, *63*, 35-47.

Herr, A. J., Jensem, M. B., Dalmay, T. & Baulcombe, D. C. (2005). RNA polymerase IV directs silencing of endogenous DNA. *Science*, *308*, 118-120.

Himber, C., Dunoyer, P., Moissiard, G., Ritzenthaler, C. & Voinnet, O. (2003). Transitivity-dependent and independent cell-to-cell movement of RNA silencing. *EMBO J.*, *22*, 4523-4533.

Hobbs, S. L. A., Kpodar, P. & DeLong, C. M. O. (1990). The effect of the T-DNA copy number, position and methylation on reporter gene expression in tobacco transformants. *Plant Mol. Biol.*, *15*, 851-864.

Howell, M. D., Fahlgren, N., Chapman, E. J., Cumbie, J. S., Sullivan, C. M., Givan, S. A., Kasschau, K. D. & Carrrington, J.C. (2007). Genome-wide analysis of the RNA-DEPENDENT RNA POLYMERACE6/DICER-LIKE4 pathway in *Arabidopsis* reveals dependency on miRNA- and tasiRNA-directed targeting. *Plant Cell*, *19*, 926-942.

Izant, J. G. & Weintraub, H. (1984). Inhibition of thymidine kinase gene expression by anti-sense RNA: a molecular approach to genetic analysis. *Cell*, *36*, 1007-1015.

Jackson, A. L., Bartz, S. R., Schelter, J., Kobayashi, S. V., Burchard, J., Mao, M., Li, B., Cavet, G. & Linsley, P. S. (2003) Expression profiling reveals off-target gene regulation by RNAi. *Nat. Biotechnol.*, *21*, 635-637.

Jones-Rhoades, M. W., Bartel, D. P. & Bartel, B. (2006). MicroRNAs and their regulatory roles in plants. *Annu. Rev. Plant Biol.*, *57*, 19-53.

Jorgensen, R. A. (1995). Cosuppression, flower color patterns, and metastable gene expression states. *Science*, *268*, 686-691.

Jorgensen, R. A., Cluster, P. D., English, J., Que, Q. & Napoli, C. A. (1996). Chalcone synthase cosuppression phenotypes in petunia flowers: comparison of sense vs. antisense constructs and single-copy vs. complex T-DNA sequences. *Plant Mol. Biol.*, *31*, 957-973.

Jorgensen, R. A., Atkinson, R. G., Forster, R. L. S. & Lucas, W. J. (1998). An RNA-based information superhighway in plants. *Science*, *279*, 1486-1487.

Kamath, R. S., Fraser, A. G., Dong, Y., Poulin, G., Durbin, R., Gotta, M., Kanapin, A., Le Bot, N., Moreno, S., Sohrmann, M., Welchman, D. P., Zipperlen, P. & Ahringer, J. (2003). Systematic functional analysis of the *Caenorhabditis elegans* genome using RNAi. *Nature*, *421*, 220-221.

Klahre, U., Crété, P., Leuenberger, S. A., Iglesias, V. A. & Meins, F., Jr. (2002). High molecular weight RNAs and small interfering RNAs induce systemic posttranscriptional gene silencing in plants. Proc. *Natl. Acad. Sci. USA,* 99, 11981-11986.

Kościańska, E., Kalantidis, K., Wypijewski, K., Sadowski, J. & Tabler, M. (2005). Analysis of RNA silencing in agroinfiltrated leaves of *Nicotiana benthamiana* and *Nicotiana tabacum*. *Plant Mol. Biol.*, *59*, 647-661.

Kurihara, Y. & Watanabe, Y. (2004). *Arabidopsis* micro-RNA biogenesis through Dicer-like 1 protein functions. *Proc. Natl. Acad. Sci. USA*, *101*, 12753-12758.

Lipardi, C., Wei, Q. & Paterson, B. M. (2001). RNAi as random degradative PCR: siRNA primers convert mRNA into dsRNAs that are degraded to generate new siRNAs. *Cell*, *107*, 297-307.

Mallory, A. C. & Vaucheret, H. (2006). Functions of microRNAs and related RNAs in plants. *Nat. Genet.*, *38*, Suppl: S31-36.

Matzke, M., Kanno, T., Huettel, B., Daxinger, L. & Matzke, A. J. (2007). Targets of RNA-directed DNA methylation. *Curr. Opin. Plant Biol.*, *10*, 512-519.

Matzke, M. A., Primig, M., Trnovsky, J. & Matzke, A. J. M. (1989). Reversible methylation and inactivation of marker genes in sequentially transformed tobacco plants. *EMBO J.*, *8*, 643-649.

Mi, S., Cai, T., Hu, Y., Chen, Y., Hodges, E., Ni, F., Wu, L., Li, S., Zhou, H., Long, C., Chen, S., Hannon, G. J. & Qi, Y. (2008) Sorting of small RNAs into *Arabidopsis* Argonaute complexes is directed by the 5' terminal nucleotide. *Cell, 133*, 116-127.

Miki, D., Itoh, R. & Shimamoto, K. (2005). RNA silencing of single and multiple members in a gene family of rice. *Plant Physiol., 138*, 1903-1913.

Mlotshwa, S., Pruss, G. J., Peragine, A., Endres, M. W., Li, J., Chen, X., Poethig, S., Bowman, L.H. & Vance, V. (2008). DIC*ER-LIKE2* plays a primary role in transitive silencing of transgenes in Arabidopsis. *PLoS ONE, 3*, e1755.

Moissiard, G., Parizotto, E. A., Himber, C. & Voinnet, O. (2008). Transitivity in *Arabidopsis* can be primed, requires the redundant action of antiviral Dicer-like 4 and Dicer-like 2, and is compromised by viral-encoded suppressor proteins. *RNA, 13*, 1268-1278.

Montgomery, T. A., Howell, M. D., Cuperus, J. T., Li, D., Hansen, J. E., Alexander, A. L., Capman, E. J., Fahlgren, N., Allen, E. & Carrington, J. C. (2008). Specificity of ARGONAUTE7-miR390 interaction and dual functionality in *TAS3 Trans*-acting siRNA formation. *Cell, 133*, 1-14.

Morel, J., Mourrain, P., Béclin, C. & Vaucheret, H. (2000). DNA methylation and chromatin structure affect transcriptional and post-transcriptional transgene silencing in Arabidopsis. *Curr. Biol., 10*, 1591-1594.

Mourrain, P., Béclin, C., Elmayan, T., Feuerbach, F., Godon, C., Morel, J., Jouette, D., Lacombe, A., Nikic, S., Picault, N., Rémoué, K., Sanial, M., Vo, T. & Vaucheret, H. (2000). *Arabidopsis SGS2* and *SGS3* genes are required for posttranscriptional gene silencing and natural virus resistance. *Cell, 101*, 533-542.

Napoli, C., Lemieux, C. & Jorgensen, R. (1990). Introduction of a chimeric chalcone synthase gene into Petunia results in reversible co-suppression of homologous genes *in trans*. *Plant Cell, 2*, 279-289.

Oka, S., Hirai, S., Choi, K. S. & Kodama, H. (2008). Preferential spreading of RNA silencing into the 3' downstream region of the transgene in tobacco. *J. Plant Biol., 51*, 227-232.

Parizotto, E. A., Dunoyer, P., Rahm, N., Himber, C. & Voinnet, O. (2004). In vivo investigation of the transcription, processing, endonucleolytic activity, and functional relevance of the spatial distribution of a plant miRNA. *Gene Dev., 18*, 2237-2242.

Peragine, A., Yoshikawa, M., Wu, G., Albrecht, H. L. & Poethig, R. S. (2004). SGS3 and SGS2/SDE1/RDR6 are required for juvenile development and the production of *trans*-acting siRNAs in Arabidopsis. *Genes Dev., 18,* 2368-2379.

Petersen, B. O. & Albrechtsen, M. (2005). Evidence implying only unprimed RdRP activity during transitive gene silencing in plants. *Plant Mol. Biol.*, 58, 575-583.

Qiu, S. & Lane, T. (2007). Implications of phase transitions in knockdown networks of transitive RNAi. IEEE Trans. *Nanobiosci.*, *6*, 68-76.

Que, Q. & Jorgensen, R. A. (1998). Homology-based control of gene expression patterns in transgenic petunia flowers. *Dev. Genet.*, *22*, 100-109.

Reddien, P. W., Bermange, A. L., Murfitt, K. J., Jennings, J. R. & Sánchez Alvarado, A. (2005). Identification of genes needed for regeneration, stem cell function, and tissue homeostasis by systematic gene perturbation in planaria. *Dev. Cell*, *8*, 635-649.

Sanders, M., Maddelein, W., Depicker, A., Montagu, M. V., Cornelissen, M. & Jacobs, J. (2002). An active role for endogenous β-1,3-glucanase genes in transgene-mediated co-suppression in tobacco. *EMBO J.*, *21*, 5824-5832.

Saxena, S., Jonsson, Z. O. & Dutta, A. (2003). Small RNAs with imperfect match to endogenous mRNA repress translation. *J. Biol. Chem.*, *278*, 44312 – 44319.

Schwach, F., Vaistij, F. E., Jones, L. & Baulcombe, D. C. (2005). An RNA-dependent RNA polymerase prevents meristem invasion by potato virus X and is required for the activity but not the production of a systemic silencing signal. *Plant Physiol.*, *138*, 1842-1852.

Shimamura, K., Oka, S., Shimotori, Y., Ohmori, T. & Kodama, H. (2007). Generation of secondary small interfering RNA in cell-autonomous and non-cell autonomous RNA silencing in tobacco. *Plant Mol. Biol.*, *63*, 803-813.

Sijen, T., Fleenor, J., Simmer, F., Thijssen, K. L., Parrish, S., Timmons, L., Plasterk, R. H. A. & Fire, A. (2001). On the role of RNA amplification in dsRNA-triggered gene silencing. *Cell*, *107*, 465-476.

Smith, N. A., Singh, S. P., Wang, M. B., Stoutjesdijk, P. A., Green, A. G. & Waterhouse, P. M. (2000) Total silencing by intron-spliced hairpin RNAs. *Nature*, *407*, 319-320.

Song, J., Smith, S. K., Hannon, G. J. & Joshua-Tor, L. (2004). Crystal structure of Argonaute and its implications fro RISC slicer activity. *Science*, *305*, 1434-1437.

Tang, G., Reinhart, B. J., Bartel, D. P. & Zamore, P. D. (2003). A biochemical framework for RNA silencing in plants. *Genes Dev.*, *17*, 49-63.

Tomita, R., Hamada, T., Horiguchi, G., Iba, K. & Kodama, H. (2004). Transgene overexpression with cognate small interfering RNA in tobacco. *FEBS Lett.*, *573*, 117-120.

Vaistij, F. E., Jones, L. & Baulcombe, D. C. (2002). Spreading of RNA targeting and DNA methylation in RNA silencing requires transcription of the target gene and a putative RNA-dependent RNA polymerase. *Plant Cell*, *14*, 857-867.

Van Der Krol, A. R., Mur, L. A., Beld, M., Mol, J. N. M. & Stuitje, A. R. (1990). Flavonoid genes in petunia: addition of a limited number of gene copies may lead to a suppression of a gene expression. *Plant Cell*, *2*, 291-299.

Van Houdt, H., Bleys, A. & Depicker, A. (2003). RNA target sequences promote spreading of RNA silencing. *Plant Physiol.*, *131*, 245-253.

Vazquez, F., Vaucheret, H., Rajagopalan, R., Lepers, C., Gasciolli, V., Mallory, A. C., Hilbert, J., Bartel, D. P. & Crété, P. (2004). Endogenous *trans*-acting siRNAs regulate the accumulation of *Arabidopsis* mRNAs. *Mol. Cell*, *16*, 69-79.

Voinnet, O., Vain, P., Angell, S. & Baulcombe, D. C. (1998). Systemic spread of sequence-specific transgene RNA degradation in plants is initiated by localized introduction of ectopic promoterless DNA. *Cell*, *95*, 177-187.

Wang, E. & Wagner, G. J. (2003). Elucidation of the functions of genes central to diterpene metabolism in tobacco trichomes using posttranscriptional gene silencing. *Planta*, *216*, 686-691.

Wassenegger, M. & Krczal, G. (2006). Nomenclature and functions of RNA-directed RNA polymerases. *Trends Plant Sci.*, *11*, 142-151.

Waterhouse, P. M., Smith, N. A. & Wang, M. B. (1999). Virus resistance and gene silencing: killing the messenger. *Trends Plant Sci.*, *4*, 452-457.

Xie, Z., Johansen, L. K., Gustafson, A. M., Kasschau, K. D., Lellis, A. D., Zilberman, D., Jacobsen, S. E. & Carrington, J. C. (2004). Genetic and functional diversification of small RNA pathways in plants. *PLoS Biol.*, *2*, 642-652.

Xu, P., Zhang, Y., Kang, L., Roossinck, M. J. & Mysore, K. S. (2006) Computational estimation and experimental verification of off-target silencing during posttranscriptional gene silencing in plants. *Plant Physiol.*, *142*, 429-440.

Yoshikawa, M., Peragine, A., Park, M. Y. & Poethig, R. S. (2005). A pathway for the biogenesis of *trans*-acting siRNAs in *Arabidopsis*. Gene Dev., *19*, 2164-2175.

Yu, B., Yang, Z., Li, J., Minakhina, S., Yang, M., Padgett, R. W., Steward, R. & Chen, X. (2005). Methylation as a crucial step in plant microRNA biogenesis. *Science*, *307*,932-935.

Zamore, P. D. (2002). Ancient pathways programmed by small RNAs. *Science*, *296*, 1265-1269.

Index

C

F

G

H

I

N

O

P

Q

R

S

T